Complex Analysis:
The Geometric Viewpoint

Second Edition

First edition published in 1990

Second edition © 2004 by
The Mathematical Association of America (Incorporated)
Library of Congress Catalog Card Number 2003114309

Complete Set ISBN 0-88385-000-1
Vol. 23, 2nd edition ISBN 0-88385-035-4

Printed in the United States of America

Current Printing (last digit):
10 9 8 7 6 5 4 3 2 1

The Carus Mathematical Monographs

Number Twenty-Three

Complex Analysis: The Geometric Viewpoint

Second Edition

Steven G. Krantz

Washington University in St. Louis

Published and Distributed by
THE MATHEMATICAL ASSOCIATION OF AMERICA

THE
CARUS MATHEMATICAL MONOGRAPHS

Published by
THE MATHEMATICAL ASSOCIATION OF AMERICA

————

Committee on Publications
Gerald L. Alexanderson, *Chair*

Editorial Board
Kenneth A. Ross, *Chair*
Joseph Auslander
Harold P. Boas
Robert E. Greene
Roger Horn
Jeffrey Lagarias
Barbara Osofsky

The following Monographs have been published:

1. *Calculus of Variations,* by G. A. Bliss (out of print)
2. *Analytic Functions of a Complex Variable,* by D. R. Curtiss (out of print)
3. *Mathematical Statistics,* by H. L. Rietz (out of print)
4. *Projective Geometry,* by J. W. Young (out of print)
5. *A History of Mathematics in America before 1900,* by D. E. Smith and Jekuthiel Ginsburg (out of print)
6. *Fourier Series and Orthogonal Polynomials,* by Dunham Jackson (out of print)
7. *Vectors and Matrices,* by C. C. MacDuffee (out of print)
8. *Rings and Ideals,* by N. H. McCoy (out of print)
9. *The Theory of Algebraic Numbers,* second edition, by Harry Pollard and Harold G. Diamond
10. *The Arithmetic Theory of Quadratic Forms,* by B. W. Jones (out of print)
11. *Irrational Numbers,* by Ivan Niven
12. *Statistical Independence in Probability, Analysis and Number Theory,* by Mark Kac
13. *A Primer of Real Functions,* third edition, by Ralph P. Boas, Jr.
14. *Combinatorial Mathematics,* by Herbert J. Ryser
15. *Noncommutative Rings,* by I. N. Herstein
16. *Dedekind Sums,* by Hans Rademacher and Emil Grosswald
17. *The Schwarz Function and Its Applications,* by Philip J. Davis
18. *Celestial Mechanics,* by Harry Pollard
19. *Field Theory and Its Classical Problems,* by Charles Robert Hadlock
20. *The Generalized Riemann Integral,* by Robert M. McLeod
21. *From Error-Correcting Codes through Sphere Packings to Simple Groups,*
    by Thomas M. Thompson
22. *Random Walks and Electric Networks,* by Peter G. Doyle and J. Laurie Snell

23. *Complex Analysis: The Geometric Viewpoint, second edition,* by Steven G. Krantz
24. *Knot Theory,* by Charles Livingston
25. *Algebra and Tiling: Homomorphisms in the Service of Geometry,* by Sherman Stein and Sándor Szabó
26. *The Sensual (Quadratic) Form,* by John H. Conway assisted by Francis Y. C. Fung
27. *A Panorama of Harmonic Analysis,* by Steven G. Krantz
28. *Inequalities from Complex Analysis,* by John P. D'Angelo
29. *Ergodic Theory of Numbers,* by Karma Dajani and Cor Kraaikamp

MAA Service Center
P. O. Box 91112
Washington, DC 20090-1112
800-331-1MAA   FAX: 301-206-9789

To my parents

Acknowledgments

This book owes its existence to many people. I thank Don Albers, Robert E. Greene, and Paul Halmos for convincing me to write it. I am grateful to Marco Abate, Harold Boas, Ralph Boas, David Drasin, Paul Halmos, Daowei Ma, David Minda, Marco Peloso, John Stapel, and Jim Walker for reading various versions of the manuscript and making valuable comments and suggestions. The Carus monograph committee of the MAA helped me to find the right level and focus for the book.

Ralph Boas, the chairman of the Carus monograph committee, played a special role in the development of this book. In addition to shepherding the project along, he provided necessary prodding and cajoling at crucial stages to keep the project on track. Paul Halmos also provided expert and sure counsel; he has been a good friend and mentor for many years. To both men I express my sincere gratitude.

For the new edition, Ken Ross as Chair of the Carus Monographs Editorial Board provided a strong and sure guiding hand. My friends Harold Boas, Daowei Ma, David Minda, Jeff McNeal, Marco Peloso, and Jim Walker offered valuable suggestions for the form of the new edition. Robert Burckel, Robert E. Greene, and John P. D'Angelo were especially generous in offering detailed commentary which led to decisive improvements. Finally, the Carus Monograph Committee provided yeoman service with many careful readings, sharp but constructive criticisms, and helpful suggestions. Surely a better book has been the result.

I thank Jim Milgram for creating TECPRINT, the world's first technical word processor. This has been a valuable tool to me for many years. I am grateful to Micki Wilderspin for performing the tedious task of converting the manuscript from TECPRINT into TEX. Beverly Joy Ruedi did a splendid job of making my project into a finished book.

A special vote of thanks goes to Robert Burckel for reading most of the manuscript inch by inch and correcting both my mathematics and my writing. His contributions have improved the quality of the book decisively. Responsibility for all remaining errors of course resides entirely with me.

—S.G.K.

Preface to the Second Edition

The warm reception with which the first edition of this book has been received has been a source of both pride and pleasure. It is a special privilege to have created a "for the record" version of Ahlfors's seminal ideas in the subject. And the geometric viewpoint continues to develop.

In the intervening decade, this author has learned a great deal more about geometric analysis, and his view of the subject has developed and broadened. It seems appropriate, therefore, to bring some new life to these pages, and to set forth a fresh enunciation of the role of curvature in basic complex function theory.

In this new edition, we explain how, in a natural and elementary manner, the hyperbolic disc is a model for the non-Euclidean geometry of Bolyai and Lobachevsky. Later on, we explain the Bergman kernel and provide an introduction to the Bergman metric.

I have many friends and colleagues to thank for their incisive remarks and suggestions about the first edition of this book. I hope that I do them justice in my efforts to implement a second edition. As always, the Mathematical Association of America has been an exemplary publisher and has provided all possible support in the publication process. I offer my humble thanks.

Preface to the First Edition

The modern geometric point of view in complex function theory began with Ahlfors's classic paper [AHL]. In that work it was demonstrated that the Schwarz lemma can be viewed as an inequality of certain differential geometric quantities on the disc (we will later learn that they are curvatures). This point of view—that substantive analytic facts can be interpreted in the language of Riemannian geometry—has developed considerably in the last fifty years. It provides new proofs of many classical results in complex analysis, and has led to new insights as well.

In this monograph we intend to introduce the reader with a standard one semester background in complex analysis to the geometric method. All geometric ideas will be developed from first principles, and only to the extent needed here. No background in geometry is assumed or required.

Chapter 0 gives a bird's eye view of classical function theory of one complex variable. We pay special attention to topics which are developed later in the book from a more advanced point of view. In this chapter we also sketch proofs of the main results, with the hope that the reader can thereby get a feeling for classical methodology before embarking on a study of the geometric method.

Chapter 1 begins a systematic treatment of the techniques of Riemannian geometry, specially tailored to the setting of one complex variable. In order that the principal ideas may be brought out most clearly, we shall concentrate on only a few themes: the Schwarz lemma, the Riemann mapping theorem, normal families, and Picard's theorems. For many readers this will be a first contact with the latter two results. The geometric method provides a particularly cogent explanation of these theorems, and can be contrasted with the more classical proofs which are discussed in Chapter 0. We shall also touch on Fatou-Julia theory, a topic which is rather technical from the analytic standpoint but completely natural from the point of view of geometry.

In Chapter 3 we introduce the Carathéodory and Kobayashi metrics, a device which is virtually unknown in the world of one complex variable. This decision allows us to introduce invariant metrics on arbitrary planar domains without resort to the uniformization theorem. We are then able to give a "differential geometric" interpretation of the Riemann mapping theorem.

The last chapter gives a brief glimpse of several complex variables. Some of the themes which were developed earlier in the book are carried over to two dimensions. Biholomorphic mappings are discussed, and the inequivalence of the ball and bidisc is proved using a geometric argument.

The language of differential geometry is not generally encountered in a first course in complex analysis. It is hoped that this volume will be used as a supplement to such a course, and that it may lead to greater familiarity with the fruitful methodology of geometry.

Contents

Acknowledgments   ix

Preface to the Second Edition   xi

Preface to the First Edition   xiii

0   Principal Ideas of Classical Function Theory   1
   1. A Glimpse of Complex Analysis..................... 1
   2. The Maximum Principle, the Schwarz Lemma, and
      Applications.. 11
   3. Normal Families and the Riemann Mapping Theorem... 16
   4. Isolated Singularities and the Theorems of Picard....... 22

1   Basic Notions of Differential Geometry   29
   0. Introductory Remarks 29
   1. Riemannian Metrics and the Concept of Length 30
   2. Calculus in the Complex Domain 38
   3. Isometries 42
   4. The Poincaré Metric............................... 45
   5. The Schwarz Lemma 55
   6. A Detour into Non-Euclidean Geometry 58

2   Curvature and Applications   67
   1.   Curvature and the Schwarz Lemma Revisited 67
   2.   Liouville's Theorem and Other Applications 73
   3.   Normal Families and the Spherical Metric 79
   4.   A Generalization of Montel's Theorem and the
       Great Picard Theorem 86

3   Some New Invariant Metrics   89
   0.   Introductory Remarks 89
   1.   The Carathéodory Metric 90
   2.   The Kobayashi Metric 93
   3.   Completeness of the Carathéodory and Kobayashi
       Metrics..103
   4.   An Application of Completeness: Automorphisms...... 121
   5.   Hyperbolicity and Curvature 133

4   Introduction to the Bergman Theory   137
   0.   Introductory Remarks 137
   1.   Bergman Basics.....................................138
   2.   Invariance Properties of the Bergman Kernel 140
   3.   Calculation of the Bergman Kernel................... 143
   4.   About the Bergman Metric 151
   5.   More on the Bergman Metric 155
   6.   Application to Conformal Mapping 156

5   A Glimpse of Several Complex Variables   161
   0.   Functions of Several Complex Variables............... 161
   1.   Basic Concepts 164
   2.   The Automorphism Groups of the Ball and Bidisc 170
   3.   Invariant Metrics and the Inequivalence of the Ball
       and the Bidisc 180

Appendix   **191**
  1.   Introduction 191
  2.   Expressing Curvature Intrinsically 191
  3.   Curvature Calculations on Planar Domains 200

Symbols   **205**

References   **209**

Index   **213**

CHAPTER **0**

Principal Ideas of Classical Function Theory

1. A Glimpse of Complex Analysis

The purpose of this book is to explain how various aspects of complex analysis can be understood both naturally and elegantly from the point of view of metric geometry. Thus, in order to set the stage for our work, we begin with a review of some of the principal ideas in complex analysis. A good companion volume for this introductory material is [GRK]. See also [BOAS] and [KR3].

Central to the subject are the Cauchy integral theorem and the Cauchy integral formula. From these follow the Cauchy estimates, Liouville's theorem, the maximum principle, Schwarz's lemma, the argument principle, Montel's theorem, and most of the other powerful and elegant results which are basic to the subject. We will discuss these results in essay form. The proofs which we provide are more conceptual than rigorous: *the aim is to depict a flow of ideas rather than absolute mathematical precision.*

We let $z \in \mathbb{C}$ denote a complex variable. If $P \in \mathbb{C}$ and $r > 0$, then we use the standard notation

$$D(P,r) = \{z \in \mathbb{C} : |z - P| < r\},$$

$$\overline{D}(P,r) = \{z \in \mathbb{C} : |z - P| \leq r\},$$

$$\partial D(P,r) = \{z \in \mathbb{C} : |z - P| = r\}.$$

We often use the lone symbol D to denote the unit disc $D(0, 1)$. A connected open set $U \subseteq \mathbb{C}$ is called a *domain*.

Complex analysis consists of the study of holomorphic functions. Let F be a complex-valued continuously differentiable function (in the sense of multivariable calculus) on a domain U in the complex plane. We write $F = u + iv$ to distinguish the real and imaginary parts of F. Then F is said to be *holomorphic*, or *analytic*, if it satisfies the Cauchy–Riemann equations:

$$\frac{\partial u}{\partial x} = \frac{\partial v}{\partial y} \quad \text{and} \quad \frac{\partial u}{\partial y} = -\frac{\partial v}{\partial x}.$$

This definition is equivalent to other familiar definitions, such as that in terms of the complex derivative, which we now discuss. If F is a function on a domain U in the complex plane and if $P \in U$, then F is said to possess a *complex derivative*, or to be complex differentiable, at P if

$$F'(P) = \frac{\partial F}{\partial z}(P) = \lim_{z \to P} \frac{F(z) - F(P)}{z - P}$$

exists. The function F is holomorphic on U if F possesses the complex derivative at each point of U.

Another useful approach to complex analytic (or holomorphic) functions is by way of power series: a function on a domain U is holomorphic if it has a convergent power series expansion $\sum_j a_j(z - P)^j$ about each point P of U.

The complex derivative definition of "holomorphic" is of great historical interest. Much effort was expended in the early days of the subject in proving that a function which is complex differentiable at each point of a domain U is in fact automatically continuously differentiable (in the usual sense of multivariable calculus), and from that point it is routine to check that the function satisfies the Cauchy–Riemann equations. The converse implication is a straightforward exercise. So, in the end, either definition is correct. From our perspective the Cauchy–Riemann equations provide the most useful point of view. This assertion will become more transparent as we develop the notion of complex integration.

Definition 1. A C^1, or *continuously differentiable,* curve in a domain $U \subseteq \mathbb{C}$ is a function $\gamma : [a, b] \to U$ from an interval in the real line into U such that γ' exists at each point of $[a, b]$ (in the one-sided sense at the endpoints) and is continuous on $[a, b]$. When there is no danger of confusion, we sometimes use the symbol γ to denote the set of points $\{\gamma(t) : t \in [a, b]\}$ as well as the function from $[a, b]$ to U.

A *piecewise continuously differentiable curve* is a single continuous curve which can be written as a finite union of continuously differentiable curves—Figure 1. A curve is called *closed* if $\gamma(a) = \gamma(b)$. It is called *simple closed* if it is closed and not self-intersecting: $\gamma(s) = \gamma(t)$ and $s \neq t$ together imply either that $s = a$ and $t = b$ or that $s = b$ and $t = a$.

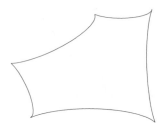

Figure 1.

A simple closed curve γ is said to be *positively oriented* if the region interior to the curve is to the left of the curve while it is being traversed from $t = a$ to $t = b$. See Figure 2. Otherwise it is called *negatively oriented*. If F is a continuous function on our open set U, then

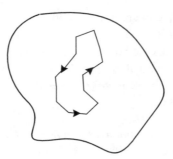

Figure 2.

we define its *complex line integral* over a continuously differentiable curve γ in U to be the quantity

$$\oint_\gamma F(z)\,dz = \int_a^b F(\gamma(t)) \cdot \gamma'(t)\,dt.$$

Here the dot $\cdot$ denotes multiplication of complex numbers.

Notice that, in analogy with the study of directed curves in Stokes's formula, the derivative of the curve is incorporated into the integral. In case γ is only piecewise continuously differentiable, we define

$$\oint_\gamma F(z)\,dz$$

by integrating along each of the continuously differentiable pieces and adding.

Now we may formulate the Cauchy integral theorem. A rigorous treatment of this result requires a discussion of deformation of curves. However, since this is only a review, we may be a bit imprecise. Let γ be a closed curve in a domain U and suppose that γ (more precisely, the *image* of γ) can be continuously deformed to a point within U. We shall call such a curve "topologically trivial (with respect to U)." In Figure 3, γ_2 is topologically trivial but γ_1 is not. We have:

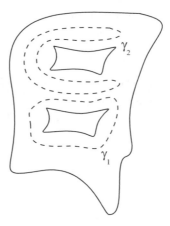

Figure 3.

Theorem 2. *Let F be a holomorphic function on a domain U and let γ be a topologically trivial, piecewise continuously differentiable, closed curve in U. Then*

$$\oint_\gamma F(z)\,dz = 0.$$

This theorem may be proved, using the Cauchy–Riemann equations, as a direct application of Stokes's theorem (see [GRK]). It will tell us, in effect, that a holomorphic function is strongly influenced on an open set by its behavior on the boundary of that set.

Now fix a point P in U and let γ be a positively oriented, topologically trivial, *simple closed curve* in U with P in its interior. Let F be holomorphic on U. By suitable limiting arguments, we may apply the Cauchy integral theorem to the function

$$G(\zeta) \equiv \begin{cases} \dfrac{F(\zeta) - F(P)}{\zeta - P} & \text{if } \zeta \neq P, \\[2mm] F'(P) & \text{if } \zeta = P. \end{cases}$$

After some calculations, the result is that

$$F(P) = \frac{1}{2\pi i} \oint_\gamma \frac{F(\zeta)}{\zeta - P}\, d\zeta.$$

This is the *Cauchy integral formula*. It shows that a holomorphic function is completely determined in the interior of γ by its behavior on the boundary curve γ itself. From this there quickly flows a wealth of information.

Theorem 3. *Let F be holomorphic on a domain U and let $P \in U$. Assume that the closed disc*

$$\overline{D}(P, r) \equiv \{z : |z - P| \le r\}$$

is contained in U. Then F may be written on $\overline{D}(P, r)$ as a convergent power series:

$$F(z) = \sum_{j=0}^{\infty} a_j (z - P)^j.$$

The convergence is absolute and uniform on $D(P, r)$.

Thus we see that, in a natural sense, holomorphic functions are generalizations of complex polynomials. The power series expansion is, in general, only local. But for many purposes this is sufficient.

Proof of Theorem 3. Observe that, for $|z - P| < r$ and $|\zeta - P| = r$, we may write

$$\frac{1}{\zeta - z} = \frac{1}{\zeta - P} \cdot \frac{1}{1 - \frac{z-P}{\zeta-P}}.$$

Since $|z - P| < r = |\zeta - P|$, we have that

$$\left| \frac{z - P}{\zeta - P} \right| < 1.$$

Thus

$$\frac{1}{\zeta - z} = \frac{1}{\zeta - P} \cdot \sum_{j=0}^{\infty} \left(\frac{z - P}{\zeta - P} \right)^j.$$

Substituting this power series expansion for the Cauchy kernel into the Cauchy integral formula on $\overline{D}(P, r)$ gives the desired power series expansion for the holomorphic function F.

∎

As an added bonus, the proof gives us a formula for the series coefficients a_j:

$$a_j = \frac{1}{2\pi i} \oint_\gamma \frac{F(\zeta)}{(\zeta - P)^{j+1}} \, d\zeta.$$

Just as in the theory of Taylor series, it turns out that the coefficients a_j must also be given by

$$a_j = \frac{1}{j!} \left(\frac{\partial^j F}{\partial z^j} \right) (P).$$

We conclude that

$$\left(\frac{\partial^j F}{\partial z^j} \right) (P) = \frac{j!}{2\pi i} \oint_\gamma \frac{F(\zeta)}{(\zeta - P)^{j+1}} \, d\zeta. \qquad (*)$$

Corollary 3.1. (Riemann removable singularities theorem). *We let $\widetilde{F}$ be a holomorphic function on a punctured disc $D'(P, r) \equiv D(P, r) \backslash \{P\}$. If $\widetilde{F}$ is bounded, then $\widetilde{F}$ continues analytically to the entire disc $D(P, r)$. That is, there is a holomorphic function F on $D(P, r)$ such that $F\big|_{D'(P,r)} = \widetilde{F}$.*

Sketch of Proof. Assume without loss of generality that $P = 0$. Consider the function $G(z)$ that is defined to equal $z^2 \cdot \widetilde{F}$ on $D'(P, r)$ and to equal 0 at $P = 0$. Then G is continuously differentiable on $D(P, r)$ and satisfies the Cauchy–Riemann equations.

The leading term of the power series expansion of G about 0 is of the form $a_2 z^2$. Thus the holomorphic function G may be divided by z^2 to define a holomorphic function F on $D(P, r)$ which agrees with $\widetilde{F}$ on $D'(P, r)$.   ∎

It is a standard fact from the theory of power series that the zeros of a function given by a power series expansion cannot accumulate in the interior of the domain of that function. Thus we have:

Theorem 4. *If F is holomorphic on a domain U, then $\{z \in U : F(z) = 0\}$ has no accumulation point in U.*

This theorem once again bears out the dictum that holomorphic functions are much like polynomials: The zero set of a polynomial $a_0 + a_1 z + a_2 z^2 + \cdots + a_n z^n$ is discrete, indeed it is finite.

The Cauchy estimates on the derivatives of a holomorphic function in terms of the supremum of the function follow from direct estimation of the formula $(*)$:

Theorem 5. *Let F be a holomorphic function on a domain U that contains the closed disc $\overline{D}(P, R)$. Let M be the supremum of $|F|$ on $\overline{D}(P, R)$. Then the derivatives of F satisfy the estimates*

$$\left| \left(\frac{\partial^j}{\partial z^j} \right) F(P) \right| \leq \frac{j! \cdot M}{R^j}.$$

An immediate corollary of the Cauchy estimates is the fact that if a sequence of holomorphic functions converges then so does the sequence of its derivatives:

Corollary 5.1. *Let $\{F_j\}$ be a sequence of holomorphic functions on a domain Ω. Suppose that the sequence converges uniformly on compact subsets of Ω. Then the sequence $\{F_j'\}$ also converges uniformly on compact subsets of Ω.*

Notice that the Cauchy estimates tell us that if F is bounded on a large disc, then its derivatives are relatively small at the center of the disc; this assertion is exploited in the next result.

Theorem 6. (Liouville). *Let F be a holomorphic function on the complex plane (an entire function) which is also bounded. Then F must be a constant.*

Proof. Assume without loss of generality that $|F|$ is bounded by 1. Fix a point P in the plane. Applying the Cauchy estimates to F on the disc $D(P, R)$ yields that

$$|F'(P)| \le \frac{1}{R^1}.$$

Letting R tend to $+\infty$ yields that $F'(P) = 0$. Since P was arbitrary, we see that $F' \equiv 0$. A simple calculus exercise now shows that F must be constant. ∎

One of the most dramatic applications of Liouville's theorem is in the proof of the fundamental theorem of algebra. That is our next task:

Theorem 7. *Let $p(z) = a_0 + a_1 z + a_2 z^2 + \cdots + a_k z^k$ be a non-constant polynomial. Then there is a point z at which p vanishes.*

Proof. Suppose not. Then $F(z) = 1/p(z)$ is an entire function. Since a non-constant polynomial blows up at infinity, F must be bounded. By Liouville's theorem, F is a constant. Hence p is constant, and therefore has degree zero. This contradiction completes the proof. ∎

Let k be the degree of the polynomial p. Notice that if the polynomial p vanishes at the point r_1, then the Euclidean algorithm implies that p is divisible by $(z - r_1)$: that is to say, $p(z) = (z - r_1) \cdot p_1(z)$ for a polynomial p_1 of degree $k - 1$. If $k - 1 \ge 1$, then we may apply the preceding result to p_1. Continuing in this fashion, we obtain that p may be expressed as a product of linear factors:

$$p(z) = (z - r_1) \cdot (z - r_2) \cdots (z - r_k).$$

We conclude this brief overview of elementary complex analysis by recalling the argument principle and Hurwitz's theorem.

Theorem 8. (The Argument Principle). *Let F be holomorphic on a domain U and let γ be a topologically trivial, positively oriented, simple closed curve in U. Assume that F does not vanish on γ. We can be sure, by Theorem 4, that there are at most finitely many, say k, zeros of F inside γ (counting multiplicity). Then we have that*

$$k = \frac{1}{2\pi i} \oint_\gamma \frac{F'(\zeta)}{F(\zeta)} \, d\zeta.$$

Sketch of Proof. By an easy reduction, it is enough to prove the result when $k = 1$. A second reduction allows us to consider the case when γ is a positively oriented circle. After a change of coordinates, let us suppose that F has a simple zero at the point $P = 0$ inside γ. By writing out the power series expansions for F and for F', we find that

$$\frac{F'(\zeta)}{F(\zeta)} = \frac{1}{\zeta} + h(\zeta),$$

where h is holomorphic near 0. Of course, h integrates to 0, by the Cauchy integral theorem. And it is easily calculated that

$$\frac{1}{2\pi i} \oint_\gamma \frac{1}{\zeta} \, d\zeta = 1,$$

completing the proof. ∎

Let U be a domain and $\{F_j\}$ a sequence of holomorphic functions on U that converges, uniformly on compact sets, to a limit function F. It is an easy consequence of the Cauchy integral formula that the limit function is also holomorphic. Let us now use the argument principle to see how the zeros of F are related to the zeros of the F_j's.

Theorem 9. (Hurwitz's Theorem). *With $\{F_j\}$ and F as above, if the F_j's are all zero-free, then either F is zero-free or F is identically zero.*

Proof. Assume that F is not identically 0. Then the zero set of F is discrete, so we can find a simple, closed, topologically trivial curve γ in U that misses the zero set completely. Therefore

$$\frac{1}{2\pi i} \oint_{\gamma} \frac{F'_j(\zeta)}{F_j(\zeta)} \, d\zeta \longrightarrow \frac{1}{2\pi i} \oint_{\gamma} \frac{F'(\zeta)}{F(\zeta)} \, d\zeta$$

as $j \to \infty$. But the expression on the left is zero for each j. Hence so is the expression on the right. So F has no zeros inside the curve γ. Since γ was chosen to be an "arbitrary" curve that misses the zeros of F, we find that F is zero-free.   ∎

2. The Maximum Principle, the Schwarz Lemma, and Applications

We begin with a brief treatment of the maximum principle. Let F be holomorphic on an open set U that contains the closed disc $\overline{D}(P, r)$. Then the Cauchy integral formula says that

$$F(P) = \frac{1}{2\pi i} \oint_{\partial D(P,r)} \frac{F(\zeta)}{\zeta - P} \, d\zeta.$$

Parametrizing the boundary of the disc by $\gamma(t) = P + re^{it}$ and writing out the definition of the line integral gives

$$F(P) = \frac{1}{2\pi} \int_0^{2\pi} F(P + re^{it}) \, dt.$$

This is the *mean value property* for a holomorphic function. Now we have:

Theorem 1. *Let F be holomorphic on a domain U. If there is a point $P \in U$ such that*

$$|F(P)| \geq |F(z)|$$

for all $z \in U$, then F is a constant function.

Sketch of Proof. Multiplying F by a constant, we may assume that $M \equiv F(P)$ is real and nonnegative. Let $S = \{z \in U : F(z) = F(P)\}$. Observe that S is nonempty since $P \in S$. Moreover, since F is a continuous function, S is trivially closed in U. To see that S is open, let $w \in S$ and suppose that $D(w, r) \subseteq U$. Now, for $0 < r' < r$, we see that

$$M = F(w) = |F(w)| = \left| \frac{1}{2\pi} \int_0^{2\pi} F(w + r'e^{it}) \, dt \right|$$

$$\leq \frac{1}{2\pi} \int_0^{2\pi} \left| F(w + r'e^{it}) \right| \, dt \leq M.$$

Since the expressions on the extreme left and right are equal, all the inequalities must be equalities. Thus, $F(w + r'e^{it}) = \left| F(w + r'e^{it}) \right| = M$ for all values of t and all $0 < r' < r$. Therefore

$$D(w, r) = \left\{ w + r'e^{it} : 0 \leq t \leq 2\pi, 0 < r' < r \right\} \subseteq S.$$

This shows that S is open.

Since S is nonempty, closed, and open, and since U is connected, it follows that $S = U$. That is what we are required to prove. ∎

For us, the main application of the maximum principle will be to the classical Schwarz lemma.

Theorem 2. (The Schwarz Lemma). *If $F : D \to D$ is holomorphic and if $F(0) = 0$, then*

$$\left| F(z) \right| \leq |z| \qquad and \qquad \left| F'(0) \right| \leq 1.$$

If either $\left| F(z) \right| = |z|$ for some $z \neq 0$ or $\left| F'(0) \right| = 1$, then F is a rotation: $F(z) \equiv e^{i\tau} \cdot z$ for some real number $\tau \in \mathbb{R}$.

Proof. The function

$$G(z) = \begin{cases} \dfrac{F(z)}{z}, & z \neq 0 \\[2mm] F'(z), & z = 0 \end{cases}$$

is holomorphic on D. Applying the maximum principle to G on the disc $\{z : |z| \leq 1 - \varepsilon\}$ for $\varepsilon > 0$ gives

$$|G(z)| \leq (1 - \varepsilon)^{-1}$$

when $|z| \leq 1 - \varepsilon$. Letting $\varepsilon \to 0^+$ yields $|G(z)| \leq 1$ on D, which is equivalent to the desired conclusion.

For uniqueness, notice that $|F(z)| = |z|$ for some $z \neq 0$ implies that $|G(z)| = 1$. The maximum principle then forces G to be a unimodular constant, hence F is a rotation. The uniqueness statement for the derivative is obtained similarly. ∎

As a simple application of the Schwarz lemma, we classify the conformal self-maps (that is, the one-to-one, onto holomorphic maps) of the unit disc to itself. We begin by claiming that if a is a complex number of modulus less than 1, then the Möbius transformation

$$\phi_a(\zeta) = \frac{\zeta - a}{1 - \overline{a}\zeta}$$

maps $D(0, 1)$ to $D(0, 1)$. [We shall use the phrase "Möbius transformation" specifically to mean a mapping of this form. Later on, we shall consider compositions of rotations with Möbius transformations.] To see this, notice that ϕ_a is certainly well defined and holomorphic in a neighborhood of $\overline{D}(0, 1)$; moreover, $\phi_a(a) = 0$. If we prove that ϕ_a maps $\partial D(0, 1)$ to $\partial D(0, 1)$, then our claim will follow from the maximum principle. But for $|\zeta| = 1$ we have that

$$|\phi_a(\zeta)| = \left| \frac{\zeta - a}{1 - \overline{a}\zeta} \right| = \left| \frac{1}{\overline{\zeta}} \cdot \frac{\zeta - a}{1 - \overline{a}\zeta} \right|$$

$$= \left| \frac{\zeta - a}{\overline{\zeta} - \overline{a}} \right| = 1,$$

since $\zeta \cdot \overline{\zeta} = 1$. This proves that $\phi_a : D(0, 1) \to D(0, 1)$. It is straightforward to check that $(\phi_a)^{-1} = \phi_{-a}$, proving that ϕ_a is one-to-one and onto.

In addition to the functions ϕ_a we also have the rotations $\rho_\tau(\zeta) \equiv e^{i\tau} \cdot \zeta$, $\tau \in \mathbb{R}$, which map the disc in a one-to-one fashion onto the disc. The remarkable fact is that these two types of maps completely characterize all conformal self-maps of the disc.

Theorem 3. *Let F be a conformal self-map of the unit disc. Then there exists a complex number a of modulus less than 1 and a real number τ such that*

$$F(\zeta) = \phi_a \circ \rho_\tau(\zeta).$$

Proof. Let $F(0) = b$ and consider the function $G \equiv \phi_b \circ F$. Then G is a conformal self-map of the unit disc and $G(0) = 0$. The Schwarz lemma tells us that $|G'(0)| \leq 1$. But we may also apply the Schwarz lemma to G^{-1} to obtain that $\left|1/G'(0)\right| = \left|(G^{-1})'(0)\right| \leq 1$. We conclude that $|G'(0)| = 1$. The uniqueness part of the Schwarz lemma then says that $G(\zeta) = \rho_\tau(\zeta)$ for some τ. But this is equivalent to

$$F = \phi_{-b} \circ \rho_\tau.$$

Setting $a = -b$ completes the proof.   ∎

Exercises. Give a similar proof to show that every conformal self-map of the disc can be written in the form $\rho_\tau \circ \phi_a$.

Verify directly that the composition of two functions of the form $\phi_a \circ \rho_\tau$ (resp. $\rho_\tau \circ \phi_a$) is also a function of that form.

It is worth noting here that the Möbius transformations ϕ_a are instances of *linear fractional transformations*. In general, a linear fractional transformation has the form

$$z \longmapsto \frac{az + b}{cz + d}$$

(with $ad - bc \neq 0$). It is elementary to see that any linear fractional transformation is a composition of translations, dilations, and inversion ($z \mapsto 1/z$). As a result, a linear fractional transformation will take circles and lines to circles and lines.

Our next step is to generalize the Schwarz lemma by removing the requirement that $F(0) = 0$. We have:

Theorem 4. (Schwarz, Pick). *If* $F : D \to D$ *is holomorphic,* $F(z_1) = w_1$ *and* $F(z_2) = w_2$, *then*

$$\left| \frac{w_1 - w_2}{1 - w_1 \overline{w}_2} \right| \leq \left| \frac{z_1 - z_2}{1 - z_1 \overline{z}_2} \right|$$

and

$$|F'(z_1)| \leq \frac{1 - |w_1|^2}{1 - |z_1|^2}.$$

If equality obtains in the first expression for some $z_1 \neq z_2$ *or if equality obtains in the second expression, then* F *must be a conformal self-map of the disc.*

Proof. Define

$$\phi(z) = \frac{z + z_1}{1 + \overline{z}_1 z}, \qquad \psi(z) = \frac{z - w_1}{1 - \overline{w}_1 z}.$$

Then $\psi \circ F \circ \phi$ satisfies the hypotheses of Schwarz's lemma. Therefore

$$\left| (\psi \circ F \circ \phi)(z) \right| \leq |z|, \qquad \forall z \in D.$$

Setting $z = \phi^{-1}(z_2)$ now gives the first inequality. Also the Schwarz lemma says that

$$\left| (\psi \circ F \circ \phi)'(0) \right| \leq 1.$$

Using the chain rule to write this out gives the second inequality.

The case of equality is analyzed as in Theorem 2. ∎

3. Normal Families and the Riemann Mapping Theorem

One of the most important concepts in topology is compactness. Compactness for a set of points in Euclidean space is, thanks to the Heine–Borel theorem, easy to understand: a set is compact if and only if it is closed and bounded. In modern times, compactness for families of functions has proved to be a powerful tool. The notion of normal families, and in particular of Montel's theorem, is historically one of the first instances of this concept. In the present section we shall treat both Montel's theorem and its application to the Riemann mapping theorem.

In an effort to keep our exposition simple, we treat here a restrictive definition of normal family; later in the book (Section 2.3), normal families will be given a more thorough treatment from a completely different point of view.

Definition 1. A family $\mathcal{F}$ of functions on a domain U will be called *normal* if every sequence in $\mathcal{F}$ has a subsequence which converges uniformly on compact subsets of U.

Now Montel's theorem says the following:

Theorem 2. (Montel). *Let U be a domain in the complex plane. Let $\mathcal{F}$ be a family of holomorphic functions on U. If there is a positive constant M such that*

$$|F(z)| \leq M$$

for all $z \in U$ and all $F \in \mathcal{F}$, then $\mathcal{F}$ is a normal family.

For the proof of this theorem, we shall need an important result from real analysis. We begin with a little terminology.

Definition 3. Let $\mathcal{F}$ be a family of functions on a common domain $S \subseteq \mathbb{R}^n$. We say that the family is *equicontinuous* if for each $\varepsilon > 0$

there is a $\delta > 0$ such that whenever $z, w \in S$ satisfy $|z - w| < \delta$, then

$$|F(z) - F(w)| < \varepsilon$$

for all $F \in \mathcal{F}$.

Notice that the property of equicontinuity is stronger than uniform continuity: not only is the choice of δ independent of the points z and w in S (i.e., it depends only on ε), but it is independent of which F from the family $\mathcal{F}$ we are considering. Next we define a companion notion.

Definition 4. Let $\mathcal{F}$ be a family of functions on a common domain $S \subseteq \mathbb{R}^n$. We say that the family is *equibounded* if there is a number $M > 0$ such that, whenever $z \in S$ and $F \in \mathcal{F}$, then

$$|F(z)| \leq M.$$

We use the notions of equicontinuity and equiboundedness to formulate the following essential result.

Proposition 5. (Ascoli/Arzelà). *Let K be a compact set in $\mathbb{R}^n$. Let $\mathcal{F}$ be an equicontinuous family of functions on K which is also equibounded. Then every sequence $\{f_j\}$ in $\mathcal{F}$ contains a subsequence $\{f_{j_k}\}$ which converges uniformly on K.*

We shall not prove the Ascoli/Arzelà theorem here. Instead we refer the interested reader to [RU1] for details. Note, however, the thrust of the theorem: if a family of rather nice objects is bounded, then it has a convergent subsequence. In other words, the theorem is a statement about sequential compactness.

Proposition 5, as stated, is about scalar-valued functions. Later on we shall have occasion to use the Ascoli/Arzelà theorem for metric-space-valued functions. The proof is just the same.

Now we apply the Ascoli/Arzelà theorem to prove Montel's theorem:

Proof of Theorem 2. Let $\overline{D}(P, R) \subseteq U$. Since U is open, the complement ^cU of U is closed. Since $\overline{D}(P, R)$ and ^cU are disjoint, there is a positive distance between the two sets: that is, there exists a $k > 0$ such that if $z \in \overline{D}(P, R)$ and $u \in {}^c U$, then $|z - u| > k$. Thus, for any $z \in \overline{D}(P, R)$ and any $F \in \mathcal{F}$ we may apply the Cauchy estimates to F on the disc $\overline{D}(z, k)$. We conclude that

$$|F'(z)| \leq \frac{M \cdot 1!}{k^1} \equiv C.$$

It is now an easy exercise, using the fact that

$$F(z) - F(w) = \int_w^z F'(\xi)\, d\xi,$$

to see that

$$\left| F(z) - F(w) \right| \leq C|z - w|$$

for any $z, w \in \overline{D}(P, R)$. This estimate shows that the family $\mathcal{F}$ is equicontinuous on $\overline{D}(P, R)$: given $\varepsilon > 0$, choose $\delta = \varepsilon/C$.

Finally, if K is any compact subset of U, then K can be covered by finitely many discs $D(P, R)$ as above, and the equicontinuity on K follows from the triangle inequality.

We may now apply the Ascoli/Arzelà theorem to $\mathcal{F}$ on K to find a uniformly convergent subsequence $\{F_{j_k}\}$ of any given sequence $\{F_j\} \subseteq \mathcal{F}$. A standard diagonalization argument then yields a subsequence $\{F_{j_{k_l}}\}$ which converges uniformly on *all* compact subsets of U. ∎

Notice that in real analysis there is no analogue for Montel's theorem. The family $\mathcal{F} = \{\sin jx\}_{j=1}^{\infty}$ is equibounded, as in the hypothesis of Montel's theorem, but there is no convergent subsequence. It is the property of being holomorphic (and hence the resulting control on first derivatives) that gives the required extra structure.

Now we turn our attention to the Riemann mapping theorem. Let U and V be open sets in the complex plane. We say that U and V are *conformally equivalent* if there is a one-to-one, onto holomorphic function (i.e., a *conformal map*)

$$\sigma : U \to V.$$

The motivation for this concept is clear: any holomorphic function F on V gives rise to a holomorphic function $F \circ \sigma$ on U and any holomorphic function G on U gives rise to a holomorphic function $G \circ \sigma^{-1}$ on V. Thus complex analysis on the two domains is, in effect, equivalent.

How can we tell when two domains are conformally equivalent? An obvious necessary condition is that they be topologically equivalent, for any conformal mapping is certainly a homeomorphism. Riemann's astonishing theorem is that if a domain U is topologically equivalent to the unit disc, and is not the entire plane, then it is conformally equivalent to the unit disc.

Theorem 6. (The Riemann Mapping Theorem). *Let U be a proper subdomain of $\mathbb{C}$ that is homeomorphic to the unit disc. Then U is conformally equivalent to the open unit disc $D(0, 1)$.*

The proof of this theorem is quite elaborate; we shall outline the principal steps. To keep matters as simple as possible, we shall assume that U is bounded.

STEP 1. Fix a point $P \in U$. Let $\mathcal{F} = \{\sigma : U \to D(0, 1)$ such that σ is one-to-one, holomorphic, and $\sigma(P) = 0\}$. We claim that $\mathcal{F}$ is nonempty. In fact, since U is bounded, there is an $R > 0$ such that $U \subseteq D(0, R)$. Then the function

$$\sigma : \zeta \mapsto \frac{1}{2R}(\zeta - P)$$

certainly maps P to 0, is holomorphic, and satisfies

$$|\sigma(\zeta)| < \frac{1}{2R}(R + R) = 1.$$

Obviously σ is one-to-one. So $\sigma \in \mathcal{F}$ and $\mathcal{F}$ is nonempty.

STEP 2. The family $\mathcal{F}$ is a normal family. In fact, the elements of $\mathcal{F}$ are holomorphic, and they are all bounded by 1. Montel's theorem thus applies, and the family is normal.

STEP 3. Let us define $M = \sup\{|\sigma'(P)| : \sigma \in \mathcal{F}\}$. We claim that M is finite. In fact, let $\overline{D}(P, r)$ be a closed disc about P that is contained in U. Then the Cauchy estimates apply: for any $\sigma \in \mathcal{F}$, we have that

$$|\sigma'(P)| \le \frac{1!}{r} = \frac{1}{r}.$$

Hence M does not exceed $1/r$.

STEP 4. We claim that there is a function $\sigma_0 \in \mathcal{F}$ such that $\sigma_0'(P) = M$. This is where normal families come in. The present step is of historical interest because Riemann asserted it without proof. Thus his theorem was doubted for several years until the step was corrected using the Dirichlet principle. Nowadays normal families are more commonly used to verify the assertion. The point is that it is fallacious to formulate an extremal problem and simply to assume that it has a solution.

To see how the argument works, we use the definition of "supremum" to conclude that there is a sequence $\{\sigma_j\}$ in $\mathcal{F}$ such that $|\sigma_j'(P)| \to M$. But since, by Step 2, $\mathcal{F}$ is a normal family, we know that there is a subsequence $\{\sigma_{j_k}\}$ that converges on compact subsets to a limit function σ_0. Now we certainly know that $\{|\sigma_{j_k}'(P)|\}$ converges to M (Corollary 5.1 of Section 1). Hence $|\sigma_0'(P)| = M$. Multiplying σ_0 by a unimodular constant now gives the function which we seek.

STEP 5. We claim that σ_0 is one-to-one on U. In fact, this follows from a clever application of the argument principle. Fix distinct points Q and R in U. Let $0 < s < |Q - R|$. Consider the functions $\psi_k(z) \equiv \sigma_{j_k}(z) - \sigma_{j_k}(Q)$ on $\overline{D}(R, s)$. Since the σ_j are all one-to-one, the functions ψ_k are all nonvanishing on $\overline{D}(R, s)$. By Hurwitz's theorem the limit function $\sigma_0(z) - \sigma_0(Q)$ is either identically zero or is nonvanishing on $\overline{D}(R, s)$. It certainly is not identically zero, for we verified in Step 4 that $\sigma_0'(P) = M > 0$. Thus $\sigma_0(z) \ne \sigma_0(Q)$ for all $z \in \overline{D}(R, s)$, in particular $\sigma_0(R) \ne \sigma_0(Q)$.

Since Q and R were arbitrary, we find that σ_0 is one-to-one.

STEP 6. We claim that σ_0 maps U *onto* the unit disc. It is in this step that the topology of U comes in. One proves, by an advanced calculus argument, that if F is any nonvanishing holomorphic function on U, then $\log F$ is a well defined holomorphic function on U. Indeed, one can define this logarithm by the formula

$$\log F(z) = \oint_{\gamma_z} \frac{F'(\zeta)}{F(\zeta)} \, d\zeta,$$

where γ_z is any piecewise C^1 path connecting P to z. The topological hypothesis on U guarantees that this definition of $\log F$ is independent of the path.

Knowing that $\log F$ exists allows us to see immediately that any power of F also exists: indeed, if $\alpha \in \mathbb{C}$, then we define

$$F^\alpha(z) = \exp\left(\alpha \log F(z)\right).$$

Now, seeking a contradiction, assume that σ_0 is not onto $D(0,1)$. Say that a point $\beta \in D(0,1)$ is omitted from the image. We now compose with certain Möbius transformations of the disc in order to normalize our mapping.

Recall the Möbius transformations ϕ_a of Section 2. Our hypothesis implies that $\phi_\beta \circ \sigma_0(\zeta)$ never vanishes on U. Thus the function

$$\mu(\zeta) = \left(\phi_\beta \circ \sigma_0(\zeta)\right)^{1/2}$$

is a well defined holomorphic function on U. Finally, we let $\tau = \mu(P)$ and set

$$\nu(\zeta) = \frac{|\mu'(P)|}{\mu'(P)} \cdot \left(\phi_\tau \circ \mu(\zeta)\right).$$

Then $\nu \in \mathcal{F}$. By definition, $\nu(P) = 0$ and a calculation shows that

$$|\nu'(P)| = \frac{1 + |\beta|}{2|\beta|^{1/2}} \cdot M > M.$$

This contradicts the choice of σ_0 as having maximal derivative at P. Thus we conclude that σ_0 is onto.

The function σ_0 is the required conformal equivalence of U with $D(0, 1)$.   ∎

We close this section by recalling Carathéodory's celebrated theorem on the boundary behavior of conformal mappings.

Theorem 7. (Carathéodory's theorem). *Let $\Omega \subseteq \mathbb{C}$ be a simply connected domain whose boundary consists of a single Jordan curve. Let $\phi : \Omega \to D$ be a conformal mapping of Ω to D (as provided by the Riemann mapping theorem). Then ϕ extends to a homeomorphism of $\overline{\Omega}$ to $\overline{D}$.*

It is interesting that Painlevé's theorem about the *smooth* continuation of conformal mappings to the boundary (which we treat in some detail in Chapter 4) actually preceded Carathéodory's theorem by several years. This historical accident actually impeded the appreciation of Painlevé's result by the ranking experts of the time.

4. Isolated Singularities and the Theorems of Picard

We begin with a brief discussion of the types of singularities which a holomorphic function can have. Let F be holomorphic on the set $U = D(P, r) \setminus \{P\}$. Then there are three possible ways that F can behave near P:

  (i) F is bounded near P;

 (ii) F is unbounded near P, but there exists a $k > 0$ such that $(z - P)^k F(z)$ is bounded near P;

(iii) neither (i) nor (ii) holds.

It is plain that (i), (ii), (iii) are mutually exclusive and cover all possibilities.

In the first instance, the Riemann removable singularities theorem guarantees that $\alpha \equiv \lim_{z \to P} F(z)$ exists and that the function

$$F(z) = \begin{cases} F(z), & z \in U \\ \alpha, & z = P \end{cases}$$

is holomorphic on $D(P, r)$.

In the second case, we see that

$$\lim_{z \to P} |F(z)| = +\infty.$$

The function F is said to have a *pole* at P. If k is the least positive integer which satisfies (ii), then it turns out that F can be written in the form

$$F(z) = (z - P)^{-k} h(z)$$

with h holomorphic on $D(P, r)$ and nonvanishing near P.

The third case is the most complex, and is the one that we wish to concentrate on here. In this circumstance F is said to have an *essential singularity* at P. We begin by proving the Casorati–Weierstrass theorem, which only begins to suggest the level of complexity involved.

Theorem 1. (Casorati–Weierstrass). *Let F be holomorphic on $U = D(P, r) \setminus \{P\}$ and suppose that F has an essential singularity at P. Then, for any $0 < s < r$, the set $\{F(z) : 0 < |z - P| < s\}$ is dense in the complex plane.*

The theorem says that holomorphic functions have a very special mode of behavior near an isolated singularity: either the function has a finite limit, or an infinite limit, or else it takes values arbitrarily close to all complex values.

Proof of Theorem 1. Suppose that the conclusion is false. Then there is a complex number β and numbers $s > 0$, $\delta > 0$ such that

$$|F(z) - \beta| > \delta, \quad \text{for all } z \in D(P, s) \setminus \{P\}.$$

Define

$$h(z) = \frac{1}{F(z) - \beta}.$$

Then $|h(z)|$ is bounded near P by the constant $1/\delta$. By the Riemann removable singularities theorem, h can be continued analytically through the point P. If $h(P) \neq 0$, then we have a contradiction since then F could be continued analytically through P. But $h(P) = 0$ would imply that F has a pole at P, and that is also false. The resulting contradiction completes the proof. ∎

It is sometimes useful to think about the point at ∞ just like any point of the finite complex plane. The map

$$z \mapsto \frac{1}{z}$$

allows us to pass back and forth between neighborhoods of 0 and neighborhoods of ∞. In particular, a function F holomorphic on a set $\{z \in \mathbb{C} : |z| > R\}$ is said to have a removable singularity, a pole, or an essential singularity at ∞ if the function $F(1/z)$ has, respectively, a removable singularity, a pole, or an essential singularity at the origin. We now record some striking properties about singularities at infinity.

Proposition 2. *If F is entire (holomorphic on all of $\mathbb{C}$) and F has a removable singularity at infinity, then F is constant.*

Sketch of Proof. By examining $F(1/z)$, we see that F must have a finite limit at ∞. Thus F is bounded. By Liouville's theorem, F is constant. ∎

Proposition 3. *If F is entire and F has a pole at ∞, then F is a polynomial.*

Sketch of Proof. Looking at $F(1/z)$, we find for some $k > 0$ that $z^k F(1/z)$ is bounded near 0. But then $z^{-k} F(z)$ is bounded near ∞.

Let P_k be the kth degree Taylor polynomial of F about zero. Then the function

$$h(z) = \frac{F(z) - P_k(z)}{z^k}$$

is holomorphic on the entire plane and bounded at ∞. Thus, by Liouville, h is constant. It follows that F is a polynomial.

∎

We learn from these two propositions that if F is entire, then F is either a polynomial or F has an essential singularity at ∞. Since the whole point of holomorphic function theory is to study a class of functions which generalizes the properties of polynomials, it is clearly important to understand essential singularities.

Picard's theorems give very sharp results on the value distribution of entire functions and, more generally, holomorphic functions near an essential singularity. We now state them.

Theorem 4. (Picard's little theorem). *If F is a non-constant entire function, then the image of F contains all complex numbers with the possible exception of one value. In other words, if an entire function omits two values, then it is constant.*

Theorem 5. (Picard's great theorem). *Let F be holomorphic on an open set $U = D(P, r) \setminus \{P\}$. If F has an essential singularity at P, then for any $0 < s < r$ we have $\{F(z) : 0 < |z - P| < s\}$ contains all complex numbers except possibly one value.*

It should now be clear that the great theorem is a generalization of the little theorem, since an entire function that is not a polynomial has an essential singularity at infinity. Let us discuss some examples to put the results in perspective.

If $F(z)$ is a non-constant polynomial and α is any complex number, then the equation $F(z) = \alpha$ always has a solution, by the fundamental theorem of algebra. Thus F takes *all* complex values. If

$F(z) = e^z$, then F takes all complex values except 0. The little theorem allows non-constant entire functions to omit one complex value. It does not allow omission of two.

Both of Picard's theorems are quite difficult to prove by classical techniques. The original proof involved the construction of a special holomorphic function from the upper half plane to $\mathbb{C} \setminus \{0, 1\}$. This function, known as the *elliptic modular function,* is a powerful tool in function theory. Here we briefly discuss the construction of the modular function and its application to the proof of the little theorem.

Begin with a conformal mapping h of the right half of the region Ω, shown in Figure 1, to the upper half plane, normalized so that $h(0) = 0$, $h(1) = 1$, $h(\infty) = \infty$. By Schwarz reflection [GRK], the map h extends to a map from the entire region Ω to $\mathbb{C} \setminus \{0, 1\}$. By more delicate reflection arguments, it is possible to continue the function analytically to a map from the entire upper half plane to $\mathbb{C} \setminus \{0, 1\}$. Denote the extended function by the symbol $\lambda(z)$. Then λ is the *elliptic modular function.*

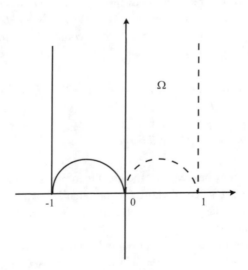

Figure 1.

Now if F is an entire function whose range omits two values, we may arrange by composition with a linear transformation for those two values to be 0 and 1. Then a function element of $\lambda^{-1} \circ F$, initially defined on a small disc, may be analytically continued to all of $\mathbb{C}$. Thus $\lambda^{-1} \circ F$ is an entire function taking values in the upper half plane. Finally, the holomorphic function

$$\rho(z) = \frac{i - z}{i + z}$$

maps the upper half plane to the unit disc. In summary, the composition $\rho \circ \lambda^{-1} \circ F$ is an entire function which takes values in the unit disc; in other words, it is bounded. By Liouville's theorem, the composition is constant. Unravelling the composition, we find that F is constant. This completes our sketch of the proof of Picard's little theorem.

The construction of the elliptic modular function is both technical and complicated. The analytic continuation argument which is required to extend an analytic function element of $\lambda^{-1} \circ F$ to all of $\mathbb{C}$ requires a great deal of theory. Other proofs of Picard's theorem arise from Nevanlinna theory, or the value distribution theory of entire functions. One of the benefits of our geometric study of complex function theory will be a rather simple and natural proof of both the little and the great Picard theorems (Section 2.4).

CHAPTER **1**
Basic Notions of Differential Geometry

0. Introductory Remarks

Differential geometry has developed into one of the most powerful mathematical tools of modern mathematics. It has become an integral part of the theories of differential equations, harmonic analysis, and complex analysis, to name just a few examples.

In spite of their importance, the techniques of geometry have not proliferated as much as they might have because of the *complexity of the language*. The characterization of differential geometry as "that portion of mathematics which is invariant under change of notation" is unfortunately rather accurate.

The best way to learn new mathematics is in the context of what is already familiar. In the milieu of one complex variable, the notions of Riemannian metric, of geodesic, and of curvature become rather simple. The standard geometric device of tensors and bundles is not necessary. The result is that one can learn the flavor and some of the methodology of differential geometry without being encumbered by its notation and machinery.

In this chapter, therefore, we shall learn some of the basic ideas of differential geometry, but *only in the complex plane*; we shall also learn a few simple applications. In later chapters we shall learn about curvature and more advanced applications.

1. Riemannian Metrics and the Concept of Length

In classical analysis a *metric* is a device for measuring distance. If X is a set then a metric λ for X is a function

$$\lambda \colon X \times X \longrightarrow \mathbb{R}$$

satisfying, for all $x, y, z \in X$,

1. $\lambda(x, y) = \lambda(y, x)$
2. $\lambda(x, y) \geq 0$ and $\lambda(x, y) = 0$ iff $x = y$;
3. $\lambda(x, y) \leq \lambda(x, z) + \gamma(z, y)$.

The trouble with a metric defined in this generality is that it does not interact well with calculus. What sort of interaction might we wish to see?

Given two points $P, Q \in X$, one would like to consider the *curve of least length* connecting P to Q. Any reasonable construction of such a curve leads to a differential equation, and thus we require that our metric lend itself to differentiation. Yet another consideration is curvature: *in the classical setting curvature is measured by the rate of change of the normal vector field.* The concepts of normal and rate of change lead inexorably to differentiation. Thus we will now take a different approach to the concept of "metric."

Recall that in multivariable calculus we define the arc length of a continuously differentiable curve $\gamma \colon [a, b] \to \mathbb{R}^2$ by the formula

$$\ell(\gamma) = \int_a^b |\dot{\gamma}(t)| \, dt. \qquad (*)$$

In point of fact, most calculus books give a heuristic which leads from a reasonable notion of length to this formula. But, rigorously speaking, $(*)$ must be considered to be the *definition* of arc length.

Notice that the definition $(*)$ hinges on the assumption that we have previously defined $|\dot{\gamma}(t)|$, the length of the tangent vector $\dot{\gamma}(t)$ to

γ at t. Riemann's insight into differential geometry was that we can define a new metric by specifying a new way to measure the lengths of tangent vectors. The most interesting examples arise when the method of measuring the length of a tangent vector varies from point to point. Thus we now give the word "metric" a new meaning; of course this new meaning will turn out to be intimately related to the older, more classical one.

Definition 1. If $\Omega \subseteq \mathbb{C}$ is a domain, then a *metric* on Ω is a continuous function $\rho(z) \geq 0$ in Ω that is twice continuously differentiable on $\{z \in \Omega : \rho(z) > 0\}$. If $z \in \Omega$ and $\xi \in \mathbb{C}$ is a vector then we define the *length* of ξ *at* z to be

$$\| \xi \|_{\rho,z} \equiv \rho(z) \cdot |\xi|,$$

where $|\xi|$ denotes the Euclidean length of the vector ξ.

Remark. Later in the book, we will consider metrics which are only continuous (or even less smooth). We want a smooth metric now because we will differentiate it to calculate curvature.   ∎

Remark. Usually our metrics ρ will be strictly positive, but it will occasionally be convenient for us to allow a metric to have isolated zeros. These will arise as zeros of holomorphic functions. The zeros of ρ should be thought of as singular points of the metric.   ∎

For the record, the metrics we are considering here are a special case of the type of differential metric called *Hermitian*. This terminology need not concern us here. Classical analysts sometimes call these metrics *conformal metrics* and write them in the form $\rho(z)|dz|$.

Technically speaking, our metric lives on the tangent bundle to the domain Ω. That is to say, the metric is a function of the variable (z, v), where v is thought of as a tangent vector at the point z. This is just a mathematical way of saying that the length of the vector v depends on the point z at which it is positioned.

Later, in Sections 4.4 and 4.5, we shall see a discussion of a different type of metric—a Riemannian metric (or Kähler metric). This will be part of our study of the Bergman kernel and metric. In that context it will be even more convenient to think of the metric as living on the tangent bundle. For most of this book, we may forego this formality.

∎

Example 1. Let Ω be any domain in $\mathbb{C}$. Define $\rho(z) = 1$ for all $z \in \Omega$. Notice that this particular metric yields that if $z \in \Omega$ and $\xi \in \mathbb{C}$ then

$$\| \xi \|_{\rho,z} \equiv \rho(z) \cdot |\xi| = |\xi|.$$

In short, this choice of metric gives the standard Euclidean notion of vector length—and this notion is independent of the base point z. This metric is usually called the *Euclidean metric*.   ∎

Example 2. Let $\Omega = \{z \in \mathbb{C}: |z| < 1\} \equiv D$, the unit disc. Let

$$\rho(z) = \frac{1}{1 - |z|^2}$$

This is the *Poincaré metric*, which has been used to gain deep insights into complex analysis on the disc. It will receive our detailed attention later on in this book. For now we do some elementary calculations with the Poincaré metric.

For any $\xi \in \mathbb{C}$ we may calculate that

$$\| \xi \|_{\rho,(1/2+0i)} = \rho(1/2 + 0i) \cdot |\xi| = \frac{4}{3} \cdot |\xi|;$$

$$\| \xi \|_{\rho,0} = \rho(0) \cdot |\xi| = 1 \cdot |\xi| = |\xi|;$$

$$\| \xi \|_{\rho,(0+.9i)} = \rho(.9i) \cdot |\xi| = \frac{1}{.19}|\xi| = (5.2631578\ldots) \cdot |\xi|. \quad ∎$$

The notion of the length of a vector varying with the base point is in contradistinction to what we learn in calculus. In calculus, a vector has direction and magnitude but *not* position. Now we declare that a vector *has position* and the way that its magnitude is calculated depends on that position. The next example shows how vectors *with position* arise in practice.

Example 3. Let $\eta(t) = t, 0 \le t \le 1/2$ and $\mu(t) = 1/2 + it, 0 \le t \le 1/2$. Let us calculate

$$\| \dot{\eta}(t) \|_{\rho,\eta(t)} \quad \text{and} \quad \| \dot{\mu}(t) \|_{\rho,\mu(t)}$$

for the Poincaré metric

$$\rho(z) = \frac{1}{1 - |z|^2}.$$

We have $\dot{\eta}(t) = 1$ for all t. Notice that the Euclidean length of $\dot{\eta}(t)$ is 1. However,

$$\| \dot{\eta}(t) \|_{\rho,\eta(t)} = \rho(\eta(t)) \cdot |\dot{\eta}(t)| = \frac{1}{1 - t^2}.$$

We also have $|\dot{\mu}(t)| = 1$ for all t and

$$\| \dot{\mu}(t) \|_{\rho,\mu(t)} = \rho(\mu(t)) \cdot |\dot{\mu}(t)| = \frac{1}{3/4 - t^2}. \quad \blacksquare$$

Tangent vectors will arise for us most often as tangents to curves. It makes sense to think of the tangent vector $\dot{\eta}(t)$ as living at $\eta(t)$ and $\dot{\mu}(t)$ as living at $\mu(t)$ for arbitrary curves η, μ. See Figure 1.

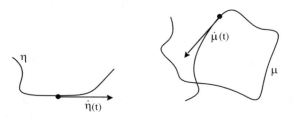

Figure 1.

Definition 2. Let $\Omega \subseteq \mathbb{C}$ be a domain and ρ a metric on Ω. If

$$\gamma : [a, b] \to \Omega$$

is a continuously differentiable curve then we define its *length in the metric* ρ to be

$$\ell_\rho(\gamma) = \int_a^b \| \dot{\gamma}(t) \|_{\rho, \gamma(t)} \ dt.$$

The length of a piecewise continuously differentiable curve is defined to be the sum of the lengths of its continuously differentiable pieces.

Notice that this definition is in complete analogy with the classical Euclidean (calculus) definition of arc length. In that more familiar setting, we use the Euclidean definition of vector length, and that in turn is the foundation for our idea of arc length. Now we have a more general means to determine the length of a vector. As a result, our integrand contains this more general notion of vector magnitude. This important idea of Riemann has led to a complete rethinking of what a geometry should be. In particular, it has led to a powerful marriage of geometric ideas with calculus ideas.

Observe also that it follows from the change of variables formula that the length of a curve is independent of its parametrization. This assertion generalizes a familiar fact from calculus. In practice, we often find it convenient to suppose that a curve being studied is parametrized with respect to arc length. This means that one unit of the parameter corresponds to one unit of length on the curve.

Classical sources write the arc length as

$$\ell_\rho(\gamma) = \int_\gamma \rho(z)|dz|,$$

but we shall not use this notation.

Example 4. Let $D \subseteq \mathbb{C}$ be the unit disc and let $\rho(z) = 1/(1 - |z|^2)$, the Poincaré metric on D. Fix $\epsilon > 0$. Let us calculate the length of the curve $\gamma(t) = t, 0 \leq t \leq 1 - \epsilon$. Now

$$\ell_\rho(\gamma) = \int_0^{1-\epsilon} \| \dot{\gamma}(t) \|_{\rho,\gamma(t)} \ dt$$

$$= \int_0^{1-\epsilon} \frac{|\dot{\gamma}(t)|}{1 - |\gamma(t)|^2} \ dt$$

$$= \int_0^{1-\epsilon} \frac{1}{1 - t^2} \ dt$$

$$= \frac{1}{2} \log \left[\frac{2 - \epsilon}{\epsilon} \right].$$

Observe that, when ϵ is small, $\ell_\rho(\gamma)$ is large. In fact

$$\lim_{\epsilon \to 0^+} \ell_\rho(\gamma) = +\infty.$$

This suggests that the boundary ∂D is infinitely far from the origin, at least along this particular path γ, in this the Poincaré metric. ∎

Remark. Notice that if we consider the metric $\rho(z) \equiv 1$, as in Example 1, then the length of a curve turns out to be the ordinary Euclidean notion of length. ∎

In a rigorous course on Riemannian geometry one proves that, for a reasonable metric (the metric needs to be *complete*), curves of least length always exist (the proper language for formulating this fact is that of geodesics, curves that have least length in an infinitesimal sense). We shall say more about completeness in Proposition 4 of Section 1.4 below. In the present concrete situation, we may bypass abstractions like this and still come up with useful information:

Example 5. Equip the disc D with the Poincaré metric $\rho(z) = 1/(1 - |z|^2)$. Fix $\epsilon > 0$. Let us prove that, among all continuously differentiable curves of the form

$$\mu(t) = t + iw(t), \qquad 0 \le t \le 1 - \epsilon,$$

that satisfy $\mu(0) = 0$ and $\mu(1 - \epsilon) = 1 - \epsilon + 0i$, the one of least length is $\gamma(t) = t$. Here $w(t)$ is a continuously differentiable, real-valued function.

In fact for any such candidate μ we have

$$
\begin{aligned}
\ell_\rho(\mu) &= \int_0^{1-\epsilon} \| \dot\mu(t) \|_{\rho,\mu(t)} \, dt \\
&= \int_0^{1-\epsilon} \frac{1}{1 - |\mu(t)|^2} |\dot\mu(t)| \, dt \\
&= \int_0^{1-\epsilon} \frac{1}{1 - t^2 - [w(t)]^2} \cdot (1 + [w'(t)]^2)^{1/2} \, dt.
\end{aligned}
$$

However,

$$
\frac{1}{1 - t^2 - [w(t)]^2} \geq \frac{1}{1 - t^2} \quad \text{and} \quad (1 + [w'(t)]^2)^{1/2} \geq 1.
$$

We conclude that

$$
\ell_\rho(\mu) \geq \int_0^{1-\epsilon} \frac{1}{1 - t^2} \, dt = \ell_\rho(\gamma).
$$

This is the desired result.

Notice that, with only small modifications, this argument can also be applied to *piecewise* continuously differentiable curves $t + i w(t)$. ∎

In fact if a piecewise continuously differentiable curve connecting the point $0 \in D$ to $(1 - \epsilon) + 0i \in D$ is *not* of the form

$$
\mu(t) = t + i w(t), \tag{$*$}
$$

then it may cross itself. Of course we can eliminate the loops and thereby create a shorter curve. If the resulting curve is still not the graph of a function, then elementary comparisons show that it will be longer than a curve of the form $(*)$ (see Figure 2). We may conclude that the curve γ in the example is the shortest of all curves connecting 0 to $(1 - \epsilon) + 0i$.

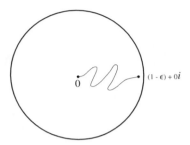

Figure 2.

The preceding discussion suggests that if a metric ρ is given on a planar domain Ω, and if P, Q are elements of Ω, then the distance in the metric ρ from P to Q should be defined as follows: Define $C_\Omega(P, Q)$ to be the collection of all piecewise continuously differentiable curves $\gamma : [0, 1] \to \Omega$ such that $\gamma(0) = P$ and $\gamma(1) = Q$. Now define the ρ-metric distance from P to Q to be

$$d_\rho(P, Q) = \inf \{\ell_\rho(\gamma) : \gamma \in C_\Omega(P, Q)\}.$$

Check for yourself that the resulting notion of distance satisfies the classical metric axioms listed at the outset of this section.

There is some subtlety connected with defining distance in this fashion. If $\rho(z) \equiv 1$, the Euclidean metric, and if Ω is the entire plane, then $d_\rho(P, Q)$ is the ordinary Euclidean distance from P to Q. But if Ω and P and Q are as shown in Figure 3, then there is no shortest curve connecting P to Q. The *distance* from P to Q is suggested by the dotted curve, but notice that this curve does not lie in Ω. The crucial issue here is whether the domain is complete in the metric, and we shall have more to say about this point later.

We may use the language and notation of distance to summarize the meaning of Examples 4 and 5 and the discussion following them: If $\rho(z) = 1/(1 - |z|^2)$ is a metric on the disc D, $P = 0$, and $Q = R + i0$, then $d_\rho(P, Q) = (1/2) \log((1 + R)/(1 - R))$.

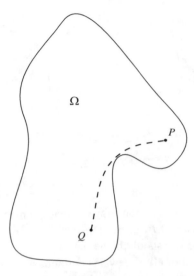

Figure 3.

2. Calculus in the Complex Domain

In order that we may be able to do calculus computations easily and efficiently in the context of complex analysis, we recast some of the basic ideas in new notation. We define the differential operators

$$\frac{\partial}{\partial z} = \frac{1}{2}\left(\frac{\partial}{\partial x} - i\frac{\partial}{\partial y}\right) \quad \text{and} \quad \frac{\partial}{\partial \overline{z}} = \frac{1}{2}\left(\frac{\partial}{\partial x} + i\frac{\partial}{\partial y}\right).$$

In complex analysis it is more convenient to use these operators than to use $\partial/\partial x$ and $\partial/\partial y$. The reason is as follows. Let $f(z) = u(z) + iv(z)$ be a complex-valued, continuously differentiable function (in the sense of multivariable calculus) on a planar domain Ω. Notice that

$$\frac{\partial}{\partial \overline{z}} f(z) = 0 \text{ on } \Omega$$

if and only if

$$\frac{1}{2}\left(\frac{\partial}{\partial x} + i\frac{\partial}{\partial y}\right)[u(z) + iv(z)] = 0 \text{ on } \Omega.$$

Taking real and imaginary parts of this last equation leads to the pair of equations

$$\frac{\partial}{\partial x}u = \frac{\partial}{\partial y}v \quad \text{and} \quad \frac{\partial}{\partial x}v = -\frac{\partial}{\partial y}u.$$

But these are just the Cauchy-Riemann equations. We conclude that

$$\frac{\partial}{\partial \bar{z}}f = 0 \quad \text{on} \quad \Omega \quad \text{iff} \quad f \quad \text{is holomorphic on } \Omega.$$

Further notice that

$$\frac{\partial}{\partial z}z = 1 \qquad \frac{\partial}{\partial z}\bar{z} = 0$$

$$\frac{\partial}{\partial \bar{z}}z = 0 \qquad \frac{\partial}{\partial \bar{z}}\bar{z} = 1.$$

Thus $\partial/\partial z$ and $\partial/\partial \bar{z}$ fit in a natural way into complex function theory. Next observe that if f is holomorphic then

$$f'(z) = \lim_{\mathbb{C} \ni h \to 0} \frac{f(z+h) - f(z)}{h}$$

$$= \lim_{\mathbb{R} \ni s \to 0} \frac{f(z+s) - f(z)}{s}$$

$$= \frac{\partial f}{\partial x}$$

$$= -i\frac{\partial f}{\partial y}$$

$$= \frac{\partial f}{\partial z}.$$

Thus we see that the notion of complex derivative, introduced in Section 0.1, is for holomorphic functions the same as $\partial f/\partial z$.

Finally observe that the Laplace operator

$$\Delta = \frac{\partial^2}{\partial x^2} + \frac{\partial^2}{\partial y^2}$$

may now be written as

$$\Delta = 4\frac{\partial}{\partial z}\frac{\partial}{\partial \overline{z}} = 4\frac{\partial}{\partial \overline{z}}\frac{\partial}{\partial z}.$$

If we are going to use complex derivatives, then we need the usual tools associated to derivatives. The linear properties (that is, the sum and scalar multiplication rules) of $\partial/\partial z$ and $\partial/\partial \overline{z}$ are obvious. The product rule is a bit more tedious, but is safely left as an exercise. However, the chain rule is more subtle; it now takes the following form:

Proposition 1. *If f and g are continuously differentiable functions, and if $f \circ g$ is well defined on some open set $U \subseteq \mathbb{C}$, then we have*

$$\frac{\partial}{\partial z}(f \circ g)(z) = \frac{\partial f}{\partial z}(g(z))\frac{\partial g}{\partial z}(z) + \frac{\partial f}{\partial \overline{z}}(g(z))\frac{\partial \overline{g}}{\partial z}(z)$$

and

$$\frac{\partial}{\partial \overline{z}}(f \circ g)(z) = \frac{\partial f}{\partial z}(g(z))\frac{\partial g}{\partial \overline{z}}(z) + \frac{\partial f}{\partial \overline{z}}(g(z))\frac{\partial \overline{g}}{\partial \overline{z}}(z).$$

Proof. We will sketch the proof of the first identity and leave the second as an exercise.

We have

$$\frac{\partial}{\partial z}(f \circ g) = \frac{1}{2}\left(\frac{\partial}{\partial x} - i\frac{\partial}{\partial y}\right)(f \circ g).$$

We write $g(z) = \alpha(z) + i\beta(z)$, with α and β real-valued functions, and apply the usual calculus chain rule for $\partial/\partial x$ and $\partial/\partial y$. We obtain that the last line equals

$$\frac{1}{2}\left(\frac{\partial f}{\partial x}\frac{\partial \alpha}{\partial x} + \frac{\partial f}{\partial y}\frac{\partial \beta}{\partial x} - i\frac{\partial f}{\partial x}\frac{\partial \alpha}{\partial y} - i\frac{\partial f}{\partial y}\frac{\partial \beta}{\partial y}\right). \tag{$*$}$$

Now, with the aid of the identities

$$\frac{\partial}{\partial x} = \frac{\partial}{\partial z} + \frac{\partial}{\partial \overline{z}} \quad \text{and} \quad \frac{\partial}{\partial y} = i\left(\frac{\partial}{\partial z} - \frac{\partial}{\partial \overline{z}}\right),$$

we may reduce the expression (∗) (after some tedious calculations) to the desired formula.   ∎

Corollary 1.1. *If either f or g is holomorphic, then*

$$\frac{\partial}{\partial z}(f \circ g)(z) = \frac{\partial f}{\partial z}(g(z))\frac{\partial g}{\partial z}(z).$$

Here is an example of the utility of our complex calculus notation.

Example 1. Let f be a nonvanishing holomorphic function on a planar domain Ω. Then

$$\Delta\bigl(\log(|f|^2)\bigr) = 0.$$

In other words, $\log(|f|^2)$ is harmonic.

   To see this, fix $P \in \Omega$ and let $U \subseteq \Omega$ be a neighborhood of P on which f has a holomorphic logarithm. Then on U we have

$$\Delta(\log(|f|^2)) = \Delta\bigl(\log f + \log \overline{f}\bigr)$$
$$= 4\frac{\partial}{\partial z}\frac{\partial}{\partial \overline{z}}\log f + 4\frac{\partial}{\partial \overline{z}}\frac{\partial}{\partial z}\log \overline{f}$$
$$= 0.$$ ∎

   We conclude this section with some exercises for the reader:

Exercise.

1. Calculate that, for $a > 0$,

$$\frac{\partial^2}{\partial z \partial \overline{z}}\log(1 + (z\overline{z})^a) = \frac{a^2(z\overline{z})^{a-1}}{(1 + (z\overline{z})^a)^2}.$$

2. If g is holomorphic, then calculate that

$$\Delta(f \circ g) = (\Delta f \circ g) \cdot |g'|^2.$$

3. If f is holomorphic, then calculate that

$$\Delta(f \circ g) = (f' \circ g) \Delta g + (f'' \circ g)[(D_x g)^2 + (D_y g)^2].$$

3. Isometries

In any mathematical subject there are morphisms: functions which preserve the relevant properties being studied. In linear algebra these are invertible linear maps, in Euclidean geometry these are rigid motions, and in Riemannian geometry these are "isometries." We now define the concept of isometry.

Definition 1. Let Ω_1 and Ω_2 be planar domains and let

$$f : \Omega_1 \to \Omega_2$$

be a continuously differentiable mapping with isolated zeros. Assume that Ω_2 is equipped with a metric ρ. We define the *pullback* of the metric ρ under the map f to be the metric on Ω_1 given by

$$f^* \rho(z) = \rho(f(z)) \cdot \left| \frac{\partial f}{\partial z} \right|.$$

Remark. The particular form that we use to define the pullback is motivated by the way that f induces mappings on tangent and cotangent vectors, but this motivation is irrelevant for us here.

It should be noted that the pullback of any metric under a conjugate holomorphic f will be the zero metric. Thus we have designed our definition of pullback so that holomorphic pullbacks will be the ones of greatest interest. This assertion will be made substantive in Proposition 3 below. ∎

Definition 2. Let Ω_1, Ω_2 be planar domains equipped with metrics ρ_1 and ρ_2, respectively. Let

$$f : \Omega_1 \to \Omega_2$$

be an onto, holomorphic mapping. If

$$f^* \rho_2(z) = \rho_1(z)$$

for all $z \in \Omega_1$ then f is called an *isometry* of the pair (Ω_1, ρ_1) with the pair (Ω_2, ρ_2).

The differential definition of isometry (Definition 2) is very natural from the point of view of differential geometry, but it is not intuitive. The next proposition relates the notion of isometry to more familiar ideas.

Proposition 3. *Let Ω_1, Ω_2 be domains and ρ_1, ρ_2 be metrics on these respective domains. If*

$$f : \Omega_1 \to \Omega_2$$

is an isometry of (Ω_1, ρ_1) to (Ω_2, ρ_2), then the following three properties hold:

(a) *If $\gamma : [a, b] \to \Omega_1$ is a continuously differentiable curve, then so is the push-forward $f_* \gamma \equiv f \circ \gamma$ and*

$$\ell_{\rho_1}(\gamma) = \ell_{\rho_2}(f_* \gamma).$$

(b) *If P, Q are elements of Ω_1 then*

$$d_{\rho_1}(P, Q) = d_{\rho_2}(f(P), f(Q)).$$

(c) *Part (b) implies that the isometry f is one-to-one. Then f^{-1} is well defined and f^{-1} is also an isometry.*

Proof. Assertion (b) is an immediate consequence of (a). Also (c) is a formal exercise in definition chasing. Therefore we shall prove (a).

By definition,

$$\ell_{\rho_2}(f_*\gamma) = \int_a^b \left\| (f_*\gamma)'(t) \right\|_{\rho_2, f_*\gamma(t)} dt$$

$$= \int_a^b \left\| \frac{\partial f}{\partial z}(\gamma(t)) \cdot \dot{\gamma}(t) \right\|_{\rho_2, f_*\gamma(t)} dt.$$

With elementary manipulations, we see that the integrand equals

$$\left| \frac{\partial f}{\partial z}(\gamma(t)) \right| \cdot \|\dot{\gamma}(t)\|_{\rho_2, f_*\gamma(t)} = \left| \frac{\partial f}{\partial z}(\gamma(t)) \right| \cdot |\dot{\gamma}(t)| \cdot \rho_2(f(\gamma(t)))$$

$$= \|\dot{\gamma}(t)\|_{f^*\rho_2, \gamma(t)}$$

$$= \|\dot{\gamma}(t)\|_{\rho_1, \gamma(t)},$$

since f is an isometry. Substituting this back into the formula for the length of $f_*\gamma$ gives

$$\ell_{\rho_2}(f_*\gamma) = \int_a^b \|\dot{\gamma}(t)\|_{\rho_1, \gamma(t)} dt = \ell_{\rho_1}(\gamma).$$

This ends the proof.   ■

It is not difficult to see that any positively oriented rigid motion of the Euclidean plane—rotations, translations, and their compositions (but *not* reflections, because they are conjugate holomorphic)—is an isometry of the Euclidean metric. In the next section we shall consider the isometries of the Poincaré metric on the disc.

If you have previously studied metric space theory or Banach space theory then you may have already encountered the term "isometry." The essential notion is that an isometry should preserve distance. In fact, you can prove as an exercise that if f is a holomorphic mapping of (Ω_1, ρ_1) onto (Ω_2, ρ_2) that preserves distance then f is an isometry according to Definition 2. (Hint: compose f with a curve and differentiate.)

We close this section by noting an important technical fact about isometries:

Proposition 4. *Let ρ_j be metrics on the domains Ω_j, $j = 1, 2, 3$, respectively. Let $f : \Omega_1 \to \Omega_2$ and $g : \Omega_2 \to \Omega_3$ be isometries. Then $g \circ f$ is an isometry of the metric ρ_1 to the metric ρ_3.*

Proof. We calculate that

$$\rho_3(g([f(z)]) \cdot |g'(f(z))| = \rho_2(f(z))$$

hence

$$\rho_3(g([f(z)]) \cdot |g'(f(z))| \cdot |f'(z)| = \rho_2(f(z)) \cdot |f'(z)| = \rho_1(z).$$

In other words,

$$\rho_3(g \circ f(z)) \cdot |(g \circ f)'(z)| = \rho_1(z),$$

as required by the definition of an isometry. ∎

4. The Poincaré Metric

The Poincaré metric on the disc has occurred in many of the examples in previous sections. This metric is the paradigm for much of what we want to do in this book, and we want to treat it in some detail here. The Poincaré metric on the disc D is given by

$$\rho(z) = \frac{1}{1 - |z|^2}.$$

(For the record, we note that there is no agreement in the literature as to what constant goes in the numerator; many references use a factor of $\sqrt{2}$ or $1/\sqrt{2}$.)

In this and succeeding sections, we shall use the phrase "conformal map" to refer to a holomorphic mapping of one planar region to another which is both one-to-one and onto.

Proposition 1. *Let ρ be the Poincaré metric on the disc D. Let $h : D \to D$ be a conformal self-map of the disc. Then h is an isometry of the pair (D, ρ) with the pair (D, ρ).*

Proof. We have that

$$h^*\rho(z) = \rho(h(z)) \cdot |h'(z)|.$$

Because of Theorem 3 of Section 0.2 (and using Proposition 4 of Section 1.3), we now have two cases:

(i) If h is a rotation, then $h(z) = \mu \cdot z$ for some unimodular constant $\mu \in \mathbb{C}$. So $|h'(z)| = 1$ and

$$h^*\rho(z) = \rho(h(z)) = \rho(\mu z) = \frac{1}{1 - |\mu z|^2} = \frac{1}{1 - |z|^2} = \rho(z)$$

as desired.

(ii) If h is a Möbius transformation (these were defined in Section 0.2, following Theorem 2), then

$$h(z) = \frac{z - a}{1 - \overline{a}z}, \qquad \text{some constant } a \in D.$$

But then

$$|h'(z)| = \frac{1 - |a|^2}{|1 - \overline{a}z|^2}$$

and

$$h^*\rho(z) = \rho\left(\frac{z - a}{1 - \overline{a}z}\right) \cdot |h'(z)|$$

$$= \frac{1}{1 - \left|\frac{z-a}{1-\overline{a}z}\right|^2} \cdot \frac{1 - |a|^2}{|1 - \overline{a}z|^2}$$

$$= \frac{1 - |a|^2}{|1 - \overline{a}z|^2 - |z - a|^2}$$

$$= \frac{1 - |a|^2}{1 - |z|^2 - |a|^2 + |a|^2|z|^2}$$

$$= \frac{1}{1 - |z|^2}$$

$$= \rho(z).$$

Since any conformal self-map of D is a composition of maps of the form (i) or (ii), the proposition is proved.   ∎

We know from this result and from Proposition 3 in Section 3 that conformal self-maps of the disc preserve Poincaré distance. To understand what this means, consider the Möbius transformation

$$\phi(z) = \frac{z + a}{1 + \bar{a}z}.$$

Then ϕ maps the disc to the disc conformally. It does not preserve Euclidean distance, but it *does* preserve Poincaré distance. See Figure 1.

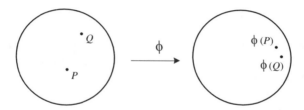

Figure 1.

We can use what we have learned so far to calculate the Poincaré metric explicitly.

Proposition 2. *If P and Q are points of the disc D, then the Poincaré distance of P to Q is*

$$d_\rho(P, Q) = \frac{1}{2} \log \left(\frac{1 + \left| \frac{P-Q}{1-\bar{P}Q} \right|}{1 - \left| \frac{P-Q}{1-\bar{P}Q} \right|} \right).$$

Proof. In case $P = 0$ and $Q = R + i0$, the result was already noted in Section 1. In the general case, note that we may define

$$\phi(z) = \frac{z - P}{1 - \bar{P}z},$$

a Möbius transformation of the disc. Then, by Proposition 1,

$$d_\rho(P, Q) = d_\rho(\phi(P), \phi(Q)) = d_\rho(0, \phi(Q)).$$

Next we have

$$d_\rho(0, \phi(Q)) = d_\rho(0, |\phi(Q)|) \qquad (*)$$

since there is a rotation of the disc taking $\phi(Q)$ to $|\phi(Q)| + i0$. Finally,

$$|\phi(Q)| = \left| \frac{P - Q}{1 - \overline{P}Q} \right|,$$

so that $(*)$ together with the special case treated in the the first sentence gives the result. ∎

One of the reasons that the Poincaré metric is so useful is that it induces the same topology as the usual Euclidean metric topology. That is our next result.

Proposition 3. *The topology induced on the disc by the Poincaré metric is the usual planar topology.*

Proof. A neighborhood basis for the topology of the Poincaré metric at the origin is given by the balls

$$\mathbf{B}(0, r) = \{z : d_\rho(0, z) < r\}.$$

However, a calculation using Proposition 2 yields that these balls are the same as the Euclidean discs

$$\left\{ z : |z| < \frac{e^{2r} - 1}{e^{2r} + 1} \right\}.$$

These discs form a neighborhood basis for the origin in the Euclidean topology. Thus we find that the two topologies are the same at the origin. Now the origin can be moved to any other point $a \in D$ by the Möbius transformation

$$z \longmapsto \frac{z + a}{1 + \overline{a}z}.$$

Since the Poincaré metric is invariant under Möbius transformations, and since Möbius transformations take circles to circles (after all, they are linear fractional), the two topologies are the same at every point.

■

One of the most striking facts about the Poincaré metric on the disc is that it turns the disc into a *complete* metric space. How could this be? The boundary is missing! The reason that the disc is complete in the Poincaré metric is the same as the reason that the plane is complete in the Euclidean metric—the boundary is infinitely far away. We now prove this assertion.

Proposition 4. *The unit disc D, when equipped with the Poincaré metric, is a complete metric space.*

Proof. Let p_j be a sequence in D that is Cauchy in the Poincaré metric. Then the sequence is bounded in that metric. So there is a positive, finite M such that

$$d_\rho(0, p_j) \le M, \quad \text{all } j.$$

With Proposition 2 this translates to

$$\frac{1}{2} \log \left(\frac{1 + |p_j|}{1 - |p_j|} \right) \le M.$$

Solving for $|p_j|$ gives

$$|p_j| \le \frac{e^{2M} - 1}{e^{2M} + 1} < 1.$$

Thus our sequence is contained in a relatively compact subset of the disc. A similar calculation yields that the sequence must in fact be Cauchy in the Euclidean metric. Therefore it converges to a limit point in the disc, as required for completeness.   ■

In the proof of Proposition 2 and in the subsequent remarks we used the fact, whose verification was sketched earlier, that the curve of least length (in the Poincaré metric) connecting 0 to a point of the form $R + i0$ is in fact a Euclidean segment. More generally, the shortest path from 0 to any point w is a rotation of the shortest path from 0 to $|w| + i0$, which is a segment. Let us now calculate the "curve of least length" connecting any two given points P and Q in the disc.

Proposition 5. *Let P, Q be elements of the unit disc. The "curve of least Poincaré length" connecting P to Q is*

$$\gamma_{P,Q}(t) = \frac{t\frac{Q-P}{1-\overline{Q}P} + P}{1 + t\overline{P} \cdot \frac{Q-P}{1-\overline{Q}P}}, \qquad 0 \le t \le 1.$$

Proof. Define the Möbius transformation

$$\phi(z) = \frac{z - P}{1 - \overline{P}z}.$$

By what we already know about shortest paths emanating from the origin, the curve $\tau(t) \equiv t \cdot \phi(Q)$ is the shortest curve from $\phi(P) = 0$ to $\phi(Q), 0 \le t \le 1$. Applying the isometry

$$\phi^{-1}(z) = \frac{z + P}{1 + \overline{P}z}$$

to τ we obtain that

$$\phi^{-1} \circ \tau(t) = \frac{(t\phi(Q) + P)}{(1 + \overline{P}t\phi(Q))}$$

is the shortest path from P to Q. Since $\phi^{-1} \circ \tau = \gamma_{P,Q}$, we are done. ∎

Let us analyze the curves discovered in Proposition 5. First notice that, since the curve $\gamma_{P,Q}$ is the image under a linear fractional transformation of a part of a line, the trace of $\gamma_{P,Q}$ is, therefore, either a line segment or an arc of a circle. In fact, if P and Q are collinear with 0, then the formula for $\gamma_{P,Q}$ quickly reduces to that for a segment; otherwise $\gamma_{P,Q}$ traces an arc of a Euclidean circle. Which circle is it?

Matters are simplest if we let t range over the entire real line and look for the whole circle. We find that $t = \infty$ corresponds to the point $1/\overline{P}$. It is now a simple, but tedious, calculation to determine the Euclidean center and radius of the circle determined by the three points $P, Q, 1/\overline{P}$ (note that, by symmetry, the circle will also pass through $1/\overline{Q}$). The circle is depicted in Figure 2. Notice that since the segment $\{t + i0 : -1 \leq t \leq 1\}$ is orthogonal to ∂D at the endpoints 1 and -1, conformality dictates that the circular arcs of least length provided by Proposition 5 are orthogonal to ∂D at the points of intersection. Some of these "geodesic arcs" are exhibited in Figure 3.

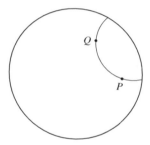

Figure 2.

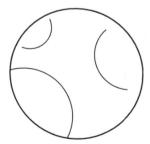

Figure 3.

A final note on this matter is that geodesics are particularly easy to calculate in the upper-half-plane realization $\mathcal{U}$ of the disc. For the geodesic circles will then have their centers on the boundary (i.e., the real line). If $\widetilde{P}$, $\widetilde{Q}$ are points of $\mathcal{U}$, then the perpendicular bisector of the segment connecting these points will intersect the real axis at the center C of the circular arc that forms the geodesic through $\widetilde{P}$ and $\widetilde{Q}$. See Figure 4.

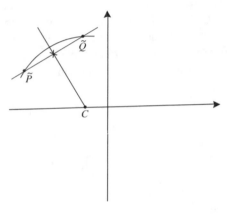

Figure 4.

We next see that the Poincaré metric is characterized by its property of invariance under conformal maps.

Proposition 6. *If $\widetilde{\rho}(z)$ is a metric on D which is such that every conformal map of the disc is an isometry of the pair $(D, \widetilde{\rho})$ with the pair $(D, \widetilde{\rho})$, then $\widetilde{\rho}$ is a constant multiple of the Poincaré metric ρ.*

Proof. The hypothesis guarantees that if $z_0 \in D$ is fixed and

$$h(z) = \frac{z + z_0}{1 + \overline{z}_0 z}$$

then

$$h^* \widetilde{\rho}(0) = \widetilde{\rho}(0).$$

Writing out the left-hand side gives

$$|h'(0)| \widetilde{\rho}(h(0)) = \widetilde{\rho}(0)$$

or

$$\widetilde{\rho}(z_0) = \frac{1}{1 - |z_0|^2} \cdot \widetilde{\rho}(0) = \widetilde{\rho}(0) \cdot \rho(z_0).$$

Thus we have exhibited $\widetilde{\rho}$ as the constant $\widetilde{\rho}(0)$ times ρ.   ∎

Now that we know that the Poincaré metric is the right metric for complex analysis on the disc, a natural next question is to determine which other maps preserve the Poincaré metric.

Proposition 7. *Let* $f : D \to D$ *be continuously differentiable and let* ρ *be the Poincaré metric. If* f *pulls back the pair* (D, ρ) *to the pair* (D, ρ), *then* f *is holomorphic and is one-to-one. We may conclude then that* f *is the composition of a Möbius transformation and a rotation (see Theorem 3 of Section 0.2).*

Proof. First suppose that $f(0) = 0$. For $R > 0$ let C_R be the set of points in D that have Poincaré distance R from 0. Since the Poincaré metric is invariant under rotations (after all, rotations are holomorphic self-maps), it follows that C_R is a Euclidean circle (however this circle will have Euclidean radius $(e^{2R} - 1)/(e^{2R} + 1)$, not R). Since $f(0) = 0$ and f preserves the metric, it follows that $f(C_R) = C_R$. The fact that f is distance-preserving then shows that f is one-to-one on each C_R. As a result, f is globally one-to-one.

Let $P \in C_R$. Then

$$\frac{|f(P) - f(0)|}{|P - 0|} = \frac{|f(P)|}{|P|} = 1.$$

Letting $R \to 0^+$ we conclude that f is conformal at the origin (that is, f preserves length infinitesimally). Since f pulls back the metric ρ, we can be sure that $\partial \rho / \partial z \neq 0$ at the origin.

Now we drop the special hypothesis that $f(0) = 0$. Pick an arbitrary $z_0 \in D$ and set $w_0 = f(z_0)$. Define

$$\phi(z) = \frac{z + z_0}{1 + \overline{z}_0 z}, \qquad \psi(z) = \frac{z - w_0}{1 - \overline{w}_0 z}.$$

Also define

$$g = \psi \circ f \circ \phi.$$

Then $g(0) = 0$ and g is an isometry (since ψ, f, and ϕ are); therefore the argument in the last paragraph applies and g is conformal at the origin. It follows as before that $\partial g / \partial z \neq 0$ at the origin.

We conclude that f is conformal at every point—since ψ and ϕ are—and $\partial f / \partial z \neq 0$ at every point. Therefore f is holomorphic.   ∎

Propositions 1 and 7 demonstrate that there are plenty of isometries of the unit disc in the Poincaré metric. On the other hand, isometries are very rigid objects. They are completely determined by their first-order behavior at just one point. While a proof of this assertion in general is beyond us at this point, we can certainly prove the result for the Poincaré metric on the disc.

Proposition 8. *Let ρ be the Poincaré metric on the disc. Let f be an isometry of the pair (D, ρ) with the pair (D, ρ). If $f(0) = 0$ and $\partial f / \partial z(0) = 1$ then $f(z) \equiv z$.*

Proof. By Proposition 7, f must be holomorphic. Since f preserves the origin and is one-to-one and onto, f must be a rotation. Since $f'(0) = 1$, it follows that f is the identity.   ∎

Corollary 8.1. *Let f and g be isometries of the pair (D, ρ) with the pair (D, ρ). Let $z_0 \in D$ and suppose that $f(z_0) = g(z_0)$ and $(\partial f / \partial z)(z_0) = (\partial g / \partial z)(z_0)$. Then $f(z) \equiv g(z)$.*

Proof. We noted in Section 2 that g^{-1} is an isometry. If ψ is a Möbius transformation that takes 0 to z_0, then $\psi^{-1} \circ g^{-1} \circ f \circ \psi$ satisfies the hypothesis of the Proposition. As a result $\psi^{-1} \circ g^{-1} \circ f \circ \psi(z) \equiv z$ or $g(z) \equiv f(z)$. ∎

5. The Schwarz Lemma

One of the important facts about the Poincaré metric is that it can be used to study not just conformal maps (as in Section 1.4) but all holomorphic maps of the disc. The key to this assertion is the classical Schwarz lemma. We begin with an elegant geometric interpretation of the Schwarz–Pick lemma (see Section 0.2).

Proposition 1. *Let $f : D \to D$ be holomorphic. Then f is distance-decreasing in the Poincaré metric ρ. That is, for any $z \in D$,*

$$f^*\rho(z) \le \rho(z).$$

The integrated form of this assertion is that if $\gamma : [0, 1] \to D$ is a continuously differentiable curve then

$$\ell_\rho(f_*\gamma) \le \ell_\rho(\gamma).$$

Therefore, if P and Q are elements of D, we may conclude that

$$d_\rho(f(P), f(Q)) \le d_\rho(P, Q).$$

Proof. Now

$$f^*\rho(z) \equiv |f'(z)|\rho(f(z)) = |f'(z)| \cdot \frac{1}{1 - |f(z)|^2}$$

and

$$\rho(z) = \frac{1}{1 - |z|^2}$$

so the asserted inequality is just the Schwarz–Pick lemma. The integrated form of the inequality now follows by the definition of ℓ_ρ. The inequality for the distance d_ρ follows from the definition of distance.

∎

Notice that, in case f is a conformal self-map of the disc, we may apply this corollary to both f and f^{-1} to conclude that f preserves Poincaré distance, giving a second proof of Proposition 1 of Section 4.

We next give an illustration of the utility of the geometric point of view. We will see that the proposition just proved gives a very elegant proof of Theorem 2 below (see [EAH] for the source of this proof).

Theorem 2. (Farkas, Ritt). *Let $f : D \to D$ be holomorphic and assume that the image $M = \{f(z) : z \in D\}$ of f has compact closure in D. Then there is a unique point $P \in D$ such that $f(P) = P$. We call P a fixed point for f.*

Proof. By hypothesis, there is an $\epsilon > 0$ such that if $m \in M$ and $|z| \geq 1$ then $|m - z| > 2\epsilon$. See Figure 1. Fix $z_0 \in D$ and define

$$g(z) = f(z) + \epsilon(f(z) - f(z_0)).$$

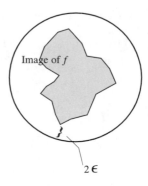

Figure 1.

Then g is holomorphic and g still maps D into D. Also

$$g'(z_0) = (1 + \epsilon) f'(z_0).$$

By the preceding proposition, g is thus distance-decreasing in the Poincaré metric. Therefore

$$g^* \rho(z_0) \le \rho(z_0).$$

Writing out the definition of g^* now yields

$$(1 + \epsilon) \cdot f^* \rho(z_0) \le \rho(z_0).$$

Note that this inequality holds for any $z_0 \in D$. But now if $\gamma : [a, b] \to D$ is any continuously differentiable curve, then we may conclude that

$$\ell_\rho(f_* \gamma) \le (1 + \epsilon)^{-1} \ell_\rho(\gamma).$$

If P and Q are elements of D and d is Poincaré distance then we have that

$$d(f(P), f(Q)) \le (1 + \epsilon)^{-1} d(P, Q).$$

We see that f is a contraction in the Poincaré metric. Recall that in Section 4 we proved that the disc D is a complete metric space when equipped with the Poincaré metric. By the contraction mapping fixed-point theorem (see [LS]), f has a unique fixed point. ∎

The next result shows how to find the fixed point, and in effect proves the contraction mapping fixed-point theorem in this special case.

Corollary 2.1. *If f is as in the theorem and P is the unique fixed point, then the iterates f, $f \circ f$, $f \circ f \circ f$, ... converge uniformly on compact sets to the constant function P.*

Proof. Let f^n denote the nth iterate of f. Let

$$\overline{\mathbf{B}}(P, R) = \{z \in D : d(z, P) \le R\}$$

be the closed Poincaré metric ball with center P and Poincaré metric radius R. Then the theorem tells us that

$$f(\overline{\mathbf{B}}(P, R)) \subseteq \overline{\mathbf{B}}(P, R/(1 + \epsilon))$$

and, more generally,

$$f^n(\overline{\mathbf{B}}(P, R)) \subseteq \overline{\mathbf{B}}(P, R/(1 + \epsilon)^n). \tag{$*$}$$

Observe that

$$\bigcup_{j=1}^{\infty} \mathbf{B}(P, j) = D.$$

Now Proposition 2 of Section 4 tells us that these non-Euclidean balls are in fact open discs in the usual Euclidean topology. Therefore every (Euclidean) compact subset K of D lies in some $\mathbf{B}(P, j)$. Then line $(*)$ implies that

$$f^n(K) \subseteq \overline{\mathbf{B}}\left(P, j/(1 + \epsilon)^n\right).$$

The result follows.   ∎

6.   A Detour into Non-Euclidean Geometry

The main purpose of this book is to show how geometry can influence and explain complex function theory. But function theory also pays back geometry in a variety of ways. Certainly the hyperbolic disc— the unit disc equipped with the Poincaré metric—is one of the most important fundamental examples in any basic course on Riemannian geometry.

     One added bonus for us is that the hyperbolic disc is also an elegant model for the classical non-Euclidean geometry of Bolyai and Lobachevsky. Since we have done all the necessary work to appreciate this construction, we take a little byway here to discuss some of the details.

The reader may recall that János Bolyai (1802–1860) and Nicolai Lobachevsky (1793–1856) are generally credited with the discovery of non-Euclidean geometry. The problem of determining the role of the parallel postulate in Euclidean geometry was two thousand years old, and many good minds had wrestled with it. Passions in the matter ran high. Bolyai's father, a provincial mathematics teacher, had worked hard on the problem. When he learned that his son, an army officer at the time, was absorbed in the problem he wrote to him

> For God's sake, I beseech you, give it up. Fear it no less than sensual passions because it, too, may take all your time, and deprive you of your health, peace of mind, and happiness.

János Bolyai did ultimately crack the problem (a couple of years in advance of Lobachevsky), producing an independent geometry in which the parallel postulate fails. The elder Bolyai conveyed the discovery to his friend Carl Friedrich Gauss (1777–1855). One can only imagine János's chagrin to read this reply from Gauss:

> If I begin with the statement that I dare not praise such a work, you will of course be startled for a moment: but I cannot do otherwise; to praise it would amount to praising myself; for the entire content of the work, the path which your son has taken, the results to which he is led, coincide almost exactly with my own meditations which have occupied my mind for from thirty to thirty-five years. On this account I find myself surprised to the extreme.
>
> My intention was, in regard to my own work, of which very little up to the present has been published, not to allow it to become known during my lifetime. Most people have not the insight to understand our conclusions and I have encountered only a few who received with any particular interest what I communicated to them. In order to understand these things, one must first have a keen perception of what is needed, and upon this point the majority are quite confused. On the other hand, it was my plan to put all down on paper eventually, so that at least it would not finally perish with me.

So I am greatly surprised to be spared this effort, and am overjoyed that it happens to be the son of my old friend who outstrips me in such a remarkable way.

Lobachevsky's publication of his results in 1829 is historically significant. In the subsequent two decades he published three additional major works on the subject. On the strength of this mathematics, Gauss recommended that Lobachevsky be given a membership in the Göttingen Scientific Society. Today non-Euclidean geometry is considered to be straightforward, and it is sometimes taught to high school students. For us, it is a convenient and charming byproduct of our construction of Poincaré geometry on the disc.

First recall the basic tenets of Euclidean geometry. There are eight *undefinable terms*: point, line, segment, parallel, right angle, the phrase "lie on," the concept of "between," and "congruent." [With some effort, the concept of "circle" can also be defined using just these undefinables.] We understand that we can obtain an intuitive understanding of these notions, but they cannot be defined rigorously (because there are no earlier concepts in terms of which to define them). The axioms of Euclidean geometry are now formulated using these undefinable terms. We in fact will give a modern formulation of these axioms (this version has been strongly influenced by the work of David Hilbert). See [KR5] for more on these ideas.

P1 Through any pair of distinct points there passes a line.

P2 For each segment $\overline{AB}$ and each segment $\overline{CD}$ there is a unique point E (on the line determined by A and B) such that B is between A and E and the segment $\overline{CD}$ is congruent to $\overline{BE}$ (Figure 1).

P3 For each point C and each point A distinct from C there exists a circle with center C and radius CA.

P4 All right angles are congruent.

These are the standard four axioms which give our Euclidean conception of geometry. The fifth axiom, a topic of intense study for two

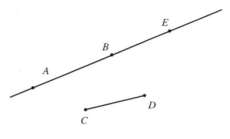

Figure 1.

thousand years, is the so-called parallel postulate (in Playfair's formulation):

P5 For each line ℓ and each point P that does not lie on ℓ there is a unique line m through P such that m is parallel to ℓ.

A *model* for this geometry is a collection of mathematical objects which can be assigned to the terms in the axiom system and which then satisfy those axioms. The most familiar context for Euclidean geometry is the familiar geometry of lines which can be drawn with a ruler and circles which can be drawn with a compass. This model satisfies all five of Euclid's axioms.

A non-Euclidean geometry is a geometry that satisfies axioms P1–P4 but not axiom P5. We now explain how the geometry of the hyperbolic disc is such a geometry.

The universe for our new geometry will be the open unit disc D in the plane. We know that the segment $\ell \equiv \{x + i0 : -1 < x < 1\}$ is a "line," in the sense that it is a geodesic in hyperbolic geometry (see Example 5 of Section 1). Now any subset of D that can be obtained as $\phi(\ell)$ for ϕ a conformal self-map of the disc will also be called a line. We will say that two of these lines are parallel if they do not intersect. Can we give an extrinsic description of these new objects which we call lines?

Recall (Theorem 3 of Section 0.2) that the conformal self-maps of the disc consist of the Möbius transformations, the rotations, and their compositions. Now it is easy to see that a rotation of ℓ will just be another diameter through the origin. More interesting is the image of ℓ under a Möbius transformation (Section 0.2).

Let us consider the specific instance of the Möbius transformation

$$\phi_{ia}(z) = \frac{z - ia}{1 - \overline{ia}z}$$

for $-1 < a < 1$ a real number. Then $\phi_{ia}(\ell)$ is the set of points

$$\left\{ \left(\frac{x(1 - a^2)}{1 + a^2 x^2}, \frac{-a(1 + x^2)}{1 + a^2 x^2} \right) : -1 < x < 1 \right\}.$$

In fact a tedious calculation shows that this is a portion of the circle of center $(0, [1+a^2]/[-2a])$ and radius $(1-a^2)/2|a|$. Of course, properly speaking, we are interested in that arc of this circle that lies inside the disc D.

By conformality, since ℓ is orthogonal to the unit circle at its endpoints, it follows that the new "line" $\phi_{ia}(\ell)$ will be orthogonal to the unit circle at its endpoints (Figure 2).

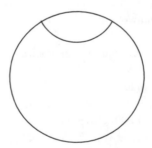

Figure 2.

Conversely, if $\mathbf{C}$ is any arc of a circle that lies in D and is perpendicular to the unit circle at its endpoints, then $\mathbf{C}$ arises as some Möbius transformation of ℓ. For, after a rotation, we may as well assume that $\mathbf{C}$ is symmetric with respect to the y-axis and lies in the upper half-plane

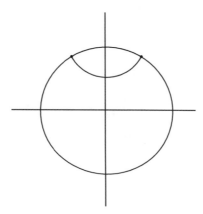

Figure 3.

(Figure 3). Then the endpoints of this arc will certainly have the form

$$\left(\frac{-(1-a^2)}{1+a^2}, \frac{-2a}{1+a^2}\right), \qquad \left(\frac{(1-a^2)}{1+a^2}, \frac{-2a}{1+a^2}\right)$$

for some real a between -1 and 0. Then we see that our circular arc **C** is just $\phi_{ia}(\ell)$.

Of course we take a *point* in our new geometry to be just a standard Euclidean point. Then it is straightforward to verify that the lines satisfy axioms P1 and P2. Axiom P3 is not relevant for our discussion of the parallel postulate since it relates to circles. Axiom P4 follows from conformality.

The punch line, of course, is that Axiom P5 is *not satisfied* by the lines in our new geometry. We repeat that two lines are parallel if they do not intersect. Now fix the line ℓ as before, and consider the point $p = (0, 1 - \epsilon)$ not on that line. There is certainly a circular arc m of the kind we have been discussing that (i) passes through p and (ii) is disjoint from ℓ (Figure 4). Now any small rotation of m will also be disjoint from ℓ and will intersect m in a point (Figure 5). Call that point of intersection q. Then q is a point that is not on the line ℓ, yet we have produced two distinct lines that pass through it and are disjoint

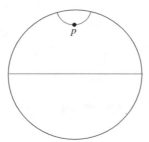

Figure 4.

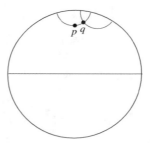

Figure 5.

from ℓ. Of course each of these two lines is, by definition, parallel to ℓ. Thus we have produced a situation that is inconsistent with the parallel postulate.

To summarize: We have a model of geometry (the classical Euclidean rectilinear model) in which all five of Euclid's axioms are true. And we have another model in which P1, P2, (P3), and P4 are true but P5 is false. This signals that Axiom P5 is *independent* of the other four.

What is elegant about this discussion is that it requires no real work: conformality, and the computations that we have already performed in another context, are sufficient to see that Euclid's 2000-year-old axiom is independent of the other four axioms of Euclid's geometry.

We close this chapter with a picture (Figure 6) of a tesselation made up of geodesics from hyperbolic geometry.

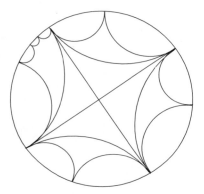

Figure 6.

CHAPTER 2

Curvature and Applications

1. Curvature and the Schwarz Lemma Revisited

If $U \subseteq \mathbb{C}$ is a planar domain and ρ is a metric on U, then the *curvature* of the metric ρ at a point $z \in U$ (at which $\rho(z) \neq 0$) is defined to be

$$\kappa_{U,\rho}(z) = \kappa(z) \equiv \frac{-\Delta \log \rho(z)}{\rho(z)^2}. \tag{$*$}$$

(Here zeros of $\rho(z)$ will result in singularities of the curvature function—κ is undefined at such points.) We will study this quantity for a bit, and then summarize and provide motivation afterward.

Since ρ is twice continuously differentiable, this definition makes sense. It assigns to each $z \in U$ a numerical quantity. The most important preliminary fact about κ is its conformal invariance:

Proposition 1. *Let U_1 and U_2 be planar domains and $h : U_1 \to U_2$ a conformal map (in particular, h' never vanishes). If ρ is a metric on U_2, then*

$$\kappa_{U_1,h^*\rho}(z) = \kappa_{U_2,\rho}(h(z)), \qquad \forall z \in U_1.$$

67

Proof. We need to calculate:

$$
\begin{aligned}
\kappa_{U_1, h^*\rho}(z) &\equiv \frac{-\Delta \log[\rho(h(z)) \cdot |h'(z)|]}{[\rho(h(z)) \cdot |h'(z)|]^2} \\
&= \frac{-\Delta \log[\rho(h(z))] - \Delta[\log |h'(z)|]}{[\rho(h(z)) \cdot |h'(z)|]^2}.
\end{aligned}
$$

Now the second term in the numerator vanishes because of Example 1 in Section 1.2. We may further simplify the numerator (use Proposition 1 in Section 1.2 as well as Exercise 2 at the end of that section) to obtain that the last line

$$
\begin{aligned}
&= \frac{-\big[(\Delta \log \rho)((h(z))\big] |h'(z)|^2}{[\rho(h(z)) \cdot |h'(z)|]^2} \\
&= \frac{-\Delta \log \rho|_{h(z)}}{\rho(h(z))^2} \\
&= \kappa_{U_2, \rho}(h(z)). \qquad \blacksquare
\end{aligned}
$$

Remark. In fact the proof gives the slightly more general fact that if U_1, U_2 are domains and $f : U_1 \rightarrow U_2$ is a holomorphic map (not necessarily one-to-one or onto), then the following holds. If ρ is a metric on U_2, then

$$
\kappa_{f^*\rho}(z) = \kappa_\rho(f(z))
$$

at every point $z \in U_1$ for which $f'(z) \neq 0$ and $\rho(f(z)) \neq 0$. Usually this is all points except for a discrete set. $\qquad \blacksquare$

From the point of view of geometry, any differential quantity which is invariant is automatically of great interest. But why do we call κ "curvature"? Why don't we call it "mass" or "humidity" or "color"? The answer is that a standard construction in Riemannian geometry assigns to any Riemannian metric a quantity called Gaussian curvature. In the setting of classical Euclidean geometry this quantity is consistent with our intuitive notion of what curvature ought to be (i.e., the rate of change of the normal). In more abstract settings, the structural equations of Cartan lead to a quantity called curvature which is invariant but which has somewhat less intuitive content. In the complex plane, the curvature calculation leads to the formula which we have used to define κ. One of the pleasant facts about the use of elementary differential geometry in complex function theory is that one does not require all the geometric machinery which leads to the formula for κ in order to make good use of κ.

Because of the considerations in the last paragraph, we shall simply accept the definition of κ as given. The geometric calculations *which lead to the definition of κ* are provided in the Appendix. Understanding of the material which follows in the text does not in any way depend on the material in this Appendix.

Let us begin our study of curvature by calculating the curvature of the Euclidean metric.

Example 1. Let U be a planar domain equipped with the Euclidean metric $\rho(z) \equiv 1$. It follows from the definition that $\kappa(z) \equiv 0$. This should be expected, for the Euclidean metric does not change from point to point. ∎

Example 2. The metric

$$\sigma(z) = \frac{2}{1 + |z|^2}$$

on $\mathbb{C}$ is called the *spherical metric* (see Section 3 below for details). A straightforward calculation shows that the curvature of σ is identically 1. Manifolds of constant curvature are a matter of great interest (see [WOL]). ∎

Next we calculate the curvature of the Poincaré metric on the disc, which calculation will be of great utility for us. We will learn that the curvature is negative. Notice that since any point of the disc may be moved to any other by a Möbius transformation, and since curvature is a conformal invariant, we will expect the curvature function to be in fact a negative constant.

Proposition 2. *Consider the disc D equipped with the Poincaré metric* ρ. *For any point* $z \in D$ *we have* $\kappa(z) = -4$.

Proof. We notice that

$$-\Delta \log \rho(z) = \Delta \log(1 - |z|^2).$$

Now we write $\Delta = 4(\partial/\partial z)(\partial/\partial \overline{z})$ (see Section 1.2) and $|z|^2 = z \cdot \overline{z}$ to obtain that this last

$$= -\frac{4}{(1 - |z|^2)^2}.$$

It follows that $\kappa(z) \equiv -4$. ∎

    It was Ahlfors [AHL1] who first realized that the Schwarz lemma is really an inequality about curvature. In his annotations to his collected works he modestly asserts that "This is an almost trivial fact and anybody who sees the need could prove it at once." However, he goes on to say (most correctly) that this point of view has been a decisive influence in modern function theory. It is a good place to begin our understanding of curvature. Here is Ahlfors's version of Schwarz's lemma:

Theorem 3. *Let the disc* $D = D(0, 1)$ *be equipped with the Poincaré metric* ρ *and let U be a planar domain equipped with a metric* σ. *Assume that, at all points of* U, σ *has curvature not exceeding* -4. *If* $f : D \rightarrow U$ *is holomorphic, then we have*

$$f^*\sigma(z) \leq \rho(z), \qquad \forall z \in D.$$

Proof following [MIS]. Let $0 < r < 1$. On the disc $D(0, r)$ define the metric

$$\rho_r(z) = \frac{r}{r^2 - |z|^2}.$$

Then straightforward calculations (or change of variable) show that ρ_r is the analogue of the Poincaré metric for $D(0, r)$: it has constant curvature -4 and is invariant under conformal maps. Define

$$v = \frac{f^*\sigma}{\rho_r}.$$

Observe that v is continuous and nonnegative on $D(0, r)$ and that $v \to 0$ when $|z| \to r$ (since $f^*\sigma$ is bounded above on $\overline{D}(0, r) \subset D$ while $\rho_r \to \infty$). It follows that v attains a maximum value $M = M(r)$ at some point $\tau \in D(0, r)$. We will show that $M \leq 1$, hence $v \leq 1$ on $D(0, r)$. Letting $r \to 1^-$ then finishes the proof.

If $f^*\sigma(\tau) = 0$, then $v \equiv 0$. So we may suppose that $f^*\sigma(\tau) > 0$. Therefore $\kappa_{f^*\sigma}$ is defined at τ. By hypothesis,

$$\kappa_{f^*\sigma} \leq -4.$$

Since $\log v$ has a maximum at τ, we have

$$0 \geq \Delta \log v(\tau) = \Delta \log f^*\sigma(\tau) - \Delta \log \rho_r(\tau)$$
$$= -\kappa_{f_*\sigma}(\tau) \cdot (f^*\sigma(\tau))^2$$
$$+ \kappa_{\rho_r}(\tau) \cdot (\rho_r(\tau))^2$$
$$\geq 4(f^*\sigma(\tau))^2 - 4(\rho_r(\tau))^2.$$

This gives

$$\frac{f^*\sigma(\tau)}{\rho_r(\tau)} \leq 1$$

or

$$M \leq 1$$

as desired. ∎

Observe that the usual Schwarz lemma is a corollary of Ahlfors's version: we take U to be the disc with σ being the Poincaré metric. Let

f be a holomorphic mapping of D to $U = D$ such that $f(0) = 0$. Then σ satisfies the hypotheses of the theorem and the conclusion is that

$$f^*\sigma(0) \leq \rho(0).$$

Unravelling the definition of $f^*\sigma$ yields

$$|f'(0)| \cdot \sigma(f(0)) \leq \rho(0).$$

But $\sigma = \rho$ and $f(0) = 0$, so this becomes

$$|f'(0)| \leq 1.$$

Exercise. The property

$$d(f(P), f(Q)) \leq d(P, Q),$$

where $d = d_\rho$ is the Poincaré distance, is called the *distance-decreasing property* of the Poincaré metric. Use this property to give a geometric proof of the other inequality in the classical Schwarz lemma.

∎

With a little more notation we can obtain a more powerful version of the Ahlfors/Schwarz lemma. Let $D(0, \alpha)$ be the open disc of radius α and center 0. For $A > 0$ define the metric ρ_α^A on $D(0, \alpha)$ by

$$\rho_\alpha^A(z) = \frac{2\alpha}{\sqrt{A}(\alpha^2 - |z|^2)}.$$

This metric has constant curvature $-A$. Then we have

Theorem 4. *Let U be a planar domain that is equipped with a metric σ whose curvature is bounded above by a negative constant $-B$. Then every holomorphic function $f : D(0, \alpha) \to U$ satisfies*

$$f^*\sigma(z) \leq \frac{\sqrt{A}}{\sqrt{B}}\rho_\alpha^A(z), \qquad \forall z \in D(0, \alpha).$$

It is a good exercise to construct a proof of this more general result, modeled of course on the proof of Theorem 3.

In the next three sections we shall see several elegant applications of Theorems 3 and 4.

2. Liouville's Theorem and Other Applications

It turns out that the invariant given by curvature provides criteria for when there do or do not exist non-constant holomorphic functions from a domain U_1 to a domain U_2. The most basic result along these lines is as follows.

Theorem 1. *Let* $U \subseteq \mathbb{C}$ *be an open set equipped with a metric* $\sigma(z)$ *having the property that its curvature* $\kappa_\sigma(z)$ *satisfies*

$$\kappa_\sigma(z) \leq -B < 0$$

for some positive constant B and for all $z \in U$. *Then any holomorphic function*

$$f : \mathbb{C} \to U$$

must be constant.

Proof. Fix $A > 0$. For $\alpha > 0$ we consider the restricted mapping

$$f : D(0, \alpha) \to U.$$

Here $D(0, \alpha)$ is the Euclidean disc with center 0 and radius α, equipped with the metric $\rho_\alpha^A(z)$ as in the last section. Theorem 4 of Section 1 yields, for any fixed z and $\alpha > |z|$, that

$$f^* \sigma(z) \leq \frac{\sqrt{A}}{\sqrt{B}} \rho_\alpha^A(z).$$

Letting $\alpha \to +\infty$ yields

$$f^*\sigma(z) \le 0,$$

hence

$$f^*\sigma(z) = 0.$$

This can only be true if $f'(z) = 0$. Since z was arbitrary, we conclude that f is constant. ∎

An immediate consequence of Theorem 1 is Liouville's theorem. Recall that an entire function is a holomorphic function whose domain is the entire complex plane.

Theorem 2. *Any bounded entire function is constant.*

Proof. Let f be such a function. After multiplying f by a constant, we may assume that the range of f lies in the unit disc. However, the Poincaré metric on the unit disc has constant curvature -4. Thus Theorem 1 applies and f must be constant.

∎

Picard's theorem (Corollary 4 below) is a dramatic strengthening of Liouville's theorem. It says that the hypothesis "bounded" may be weakened considerably, yet the same conclusion may be drawn.

Let us begin our discussion with a trivial example.

Example 1. Let f be an entire function taking values in the set

$$S = \mathbb{C} \setminus \{x + i0 : 0 \le x \le 1\}.$$

Following f by the mapping

$$\phi(z) = \frac{z}{z-1},$$

we obtain an entire function $g = \phi \circ f$ taking values in $\mathbb{C}$ less the set $\{x + i0 : x \leq 0\}$. If $r(z)$ is the principal branch of the square root function, then $h(z) = r \circ g(z)$ is an entire function taking values in the right half plane. Now the Cayley map

$$c(z) = \frac{z - 1}{z + 1}$$

takes the right half plane to the unit disc. So $u(z) = c \circ h(z)$ is a bounded entire function. We conclude from Theorem 1 that u is constant. Unravelling our construction, we have that f is constant. ∎

The point of this easy example is that, far from being bounded, an entire function need only omit a segment from its values in order that it be forced to be constant. And a small modification of the proof shows that the segment can be arbitrarily short. Picard considered the question of how large a set can be omitted from the image of a non-constant entire function.

Let us pursue the same line of inquiry rather modestly by asking whether a non-constant entire function can omit one complex value. The answer is "yes," for $f(z) = e^z$ assumes all complex values except zero. It also turns out (see Section 3.5) that it is impossible to construct a metric on the plane less a point that has negative curvature bounded away from zero.

The next step is to ask whether a non-constant entire function f can omit two values. The striking answer, discovered by Picard, is "no." Because of Theorem 1, it suffices for us to prove the following:

Theorem 3. *Let U be a planar open set such that $\mathbb{C} \setminus U$ contains at least two points. Then U admits a metric μ whose curvature $\kappa_\mu(z)$ satisfies*

$$\kappa_\mu(z) \leq -B < 0$$

for some positive constant B.

Proof. After applying an invertible linear map to U we may take the two omitted points to be 0 and 1 (if there are more than two omitted points, we may ignore the extras). Thus we will construct a metric of strictly negative curvature on $\mathbb{C}_{0,1} \equiv \mathbb{C} \setminus \{0, 1\}$.

Define

$$\mu(z) = \left[\frac{\left(1 + |z|^{1/3}\right)^{1/2}}{|z|^{5/6}} \right] \cdot \left[\frac{\left(1 + |z - 1|^{1/3}\right)^{1/2}}{|z - 1|^{5/6}} \right].$$

(After the proof, we will discuss where this nonintuitive definition came from.) Of course μ is positive and smooth on $\mathbb{C}_{0,1}$. We proceed to calculate the curvature of μ.

First notice that, away from the origin,

$$\Delta \left(\log |z|^{5/6} \right) = \frac{5}{12} \Delta \left(\log |z|^2 \right) = 0$$

by Example 1 of Section 1.2. Thus

$$\Delta \log \left[\frac{\left(1 + |z|^{1/3}\right)^{1/2}}{|z|^{5/6}} \right] = \frac{1}{2} \Delta \log \left(1 + |z|^{1/3} \right)$$

$$= 2 \frac{\partial^2}{\partial z \partial \bar{z}} \log \left(1 + [z \cdot \bar{z}]^{1/6} \right).$$

Now a straightforward calculation (use the Exercise at the end of Section 1.2) leads to the identity

$$\Delta \log \left[\frac{\left(1 + |z|^{1/3}\right)^{1/2}}{|z|^{5/6}} \right] = \frac{1}{18} \frac{1}{|z|^{5/3} \left(1 + |z|^{1/3}\right)^2}.$$

The very same calculation shows that

$$\Delta \log \left[\frac{\left(1 + |z - 1|^{1/3}\right)^{1/2}}{|z - 1|^{5/6}} \right] = \frac{1}{18} \frac{1}{|z - 1|^{5/3} \left(1 + |z - 1|^{1/3}\right)^2}.$$

The definition of curvature now yields that

$$
\kappa_\mu(z) = -\frac{1}{18}\left[\frac{|z-1|^{5/3}}{\left(1+|z|^{1/3}\right)^3\left(1+|z-1|^{1/3}\right)}\right.
$$

$$
\left.+\frac{|z|^{5/3}}{\left(1+|z|^{1/3}\right)\left(1+|z-1|^{1/3}\right)^3}\right].
$$

We record the following obvious facts:

(a) $\kappa_\mu(z) < 0$ for all $z \in \mathbb{C}_{0,1}$;

(b) $\lim_{z\to 0}\kappa_\mu(z) = -\dfrac{1}{36}$;

(c) $\lim_{z\to 1}\kappa_\mu(z) = -\dfrac{1}{36}$;

(d) $\lim_{z\to\infty}\kappa_\mu(z) = -\infty$.

It follows immediately that κ_μ is bounded from above by a negative constant.   ∎

Remark. Now we discuss the motivation for the construction of μ. When you reach the discussion in Section 3.5, you will get a thorough mathematical treatment of these issues.

On looking at the definition of μ, one sees that the first factor is singular at 0 and the second is singular at 1. Let us concentrate on the first of these.

Since the expression defining curvature is rotationally invariant, it is plausible that the metric we define would also be rotationally invariant about its singularities. Thus it should be a function of $|z|$. Hence one would like to choose exponents α, β so that $(1+|z|^\alpha)^\beta$ defines a metric of negative curvature. However, a calculation reveals that the α, β which are suitable for z large are not suitable for z small and vice-versa. This explains why the expression has powers both of $|z|$ (for behavior near 0) and of $(1+|z|)$ (for behavior near ∞). A similar discussion applies to the factors $|z-1|^\alpha$.

Our calculations thus lead us to design the metric so that it behaves like $|z|^{-4/3}$ near ∞ and behaves like $|z|^{-5/6}$ (respectively, $|z-1|^{-5/6}$) near 0 (respectively, 1).

Of course there is more going on than these brief comments might indicate. For the construction is not possible on $\mathbb{C} \setminus \{0\}$. One *must* work on a plane with two punctures. There is a delicate balance between the factors (discussed above) which are centered at 0 and those which are centered at 1. ∎

We formulate Picard's little theorem as a corollary of Theorem 1 and Theorem 3:

Corollary 4. (Picard). *Let f be an entire analytic function taking values in a set U. If $\mathbb{C} \setminus U$ contains at least two points, then f is constant.*

Proof. Since $\mathbb{C} \setminus U$ contains at least two points, Theorem 3 says that there is a metric of strictly negative curvature on U. But then Theorem 1 implies that any entire function taking values in U is constant. ∎

Entire functions are of two types: there are polynomials and non-polynomials (*transcendental* entire functions). Notice that a polynomial has a pole at infinity. Conversely, any entire function with a pole at infinity is a polynomial (see Section 0.4). So a transcendental entire function cannot have a pole at infinity and, being unbounded (by Liouville), cannot have a removable singularity at infinity. It therefore has an essential singularity at infinity.

Now notice that, by the fundamental theorem of algebra, a polynomial assumes all complex values. For a transcendental function, we analyze its essential singularity at infinity by recalling the Casorati-Weierstrass theorem (see Section 0.4): if 0 is an essential singularity for a holomorphic function f on a punctured disc $D'(0, \epsilon) \equiv D(0, \epsilon) \setminus \{0\}$ then f assumes values on $D'(0, \epsilon)$ which are *dense* in the complex plane. One might therefore conjecture that the essential feature of Picard's theorem is not that the function being considered is entire, but rather that it has an essential singularity at infinity. This is indeed the case and is the content of the great Picard theorem, which we treat in Section 4.

3. Normal Families and the Spherical Metric

Recall that in Section 0.3 we briefly considered the concept of normal family. Now we go into the subject more deeply; in particular, we give a more general definition of the concept of normal family. We shall learn that the property of being normal is, when suitably formulated, equivalent to an equicontinuity condition formulated in certain metrics. We begin with our new definition:

Definition 1. Let Ω be a planar domain and let $\{g_j\}$ be a sequence of complex-valued functions on Ω. We say that the sequence is *normally convergent* to a limit function g if, for every $\epsilon > 0$ and every compact set $K \subseteq \Omega$, there is a finite, positive J such that whenever $j > J$ and $z \in K$, then

$$|g_j(z) - g(z)| < \epsilon.$$

In short, the sequence is normally convergent if it converges uniformly on compact subsets of Ω.

Definition 2. Let Ω be a planar domain and let $\{g_j\}$ be a sequence of complex-valued functions on Ω. We say that the sequence is *compactly divergent* if, for every compact set $K \subseteq \Omega$ and every compact set $L \subseteq \mathbb{C}$, there is a finite, positive J such that if $j > J$ and $z \in K$ then $g_j(z) \notin L$.

Briefly, the sequence is compactly divergent if it converges to ∞, uniformly on compact sets.

Definition 3. Let Ω be a planar domain and let $\mathcal{F}$ be a family of complex-valued functions on Ω. We call $\mathcal{F}$ a *normal family* if every sequence of elements of $\mathcal{F}$ has either a subsequence which is normally convergent or a subsequence which is compactly divergent.

Example 1. Let $\mathcal{F} = \{f_j\}$, with $f_j(z) = z^j$, $j = 1, 2, \ldots$. Then $\mathcal{F}$ is a normal family on $U_1 = \{z : |z| < 1\}$ because every subsequence converges normally to the zero function. Also $\mathcal{F}$ is a normal family on $U_2 = \{z : |z| > 1\}$ because every subsequence is compactly divergent.

However, $\mathcal{F}$ is not normal on any domain U_3 which contains a point of the unit circle. The reason is that such a U_3 will contain points inside the circle, on which $\{f_j\}$ converges, and points outside the circle, on which $\{f_j\}$ diverges.   ∎

Notice that, contrary to many complex analysis texts, we do not require in our definition of normal family that the functions in question be holomorphic. In point of fact, normal families are useful outside of complex function theory and it is helpful to have some extra flexibility. However, if we do assume in advance that the functions we are treating are holomorphic, then there are some beautiful theorems about normal families, as we shall now see.

The standard result in the theory of normal families of holomorphic functions is Montel's theorem:

Theorem 4. *Let Ω be a planar domain and let $\mathcal{F}$ be a family of holomorphic functions on Ω. If for each compact $K \subseteq \Omega$ there is a number M_K such that*

$$|f(z)| \leq M_K, \qquad \forall z \in K, f \in \mathcal{F}, \qquad (*)$$

then $\mathcal{F}$ is a normal family.

Corollary 4.1. *If Ω, $\mathcal{F}$ are as in the theorem and if instead of $(*)$ we have a single constant M such that*

$$|f(z)| \leq M, \qquad \forall z \in \Omega, f \in \mathcal{F},$$

then $\mathcal{F}$ is a normal family.

Remark. Notice that no family $\mathcal{F}$ of holomorphic functions satisfying the hypotheses of Montel's theorem could have a compactly divergent subsequence. Thus such a family $\mathcal{F}$ has the property that every sequence of elements of $\mathcal{F}$ has a subsequence that is normally convergent.

After we apply some geometry to the study of normal families, we will be able to derive a version of Montel's theorem which is much more general, and which is invariant in a natural sense; it will allow for compact divergence as well. ∎

The proof of Montel's theorem is discussed in Section 0.3; further details may be found, for instance, in [AHL2, p. 219, ff.] or [GRK]. Now we turn to some examples.

Example 2. Let $\mathcal{F}$ be the family of all holomorphic functions on a domain Ω that take values in the right half plane. Define

$$\phi(z) = \frac{z-1}{z+1}.$$

Then $\mathcal{G} \equiv \{\phi \circ f : f \in \mathcal{F}\}$ is a family of holomorphic functions taking values in the unit disc. By the corollary to Montel's theorem, $\mathcal{G}$ is a normal family. It follows that $\mathcal{F}$ is a normal family. ∎

Example 3. Let

$$U_0 = \mathbb{C} \setminus \{x + i0 : 0 \le x \le 1\}.$$

Let $\mathcal{F}$ be all those functions holomorphic in a given domain U and taking values in U_0. We claim that $\mathcal{F}$ is a normal family.

In fact, in Example 1 of Section 2 we found a univalent (one-to-one) holomorphic map, call it μ, of U_0 into the unit disc. Therefore $\mathcal{G} = \{\mu \circ f : f \in \mathcal{F}\}$ is a normal family. It follows that $\mathcal{F}$ is a normal family. ∎

With these examples we are trying to suggest that the concept of normal families is related to Liouville's theorem and Picard's theorem (this notion is sometimes called "Bloch's principle"). In fact this is true, and we can now begin to see the assertion by looking at matters geometrically. We begin with the concept of the spherical metric.

The stereographic projection p gives a map from the complex plane $\mathbb{C}$ to the Riemann sphere $\hat{\mathbb{C}}$ (see [AHL2, p. 19]). This map is illustrated in Figure 1, with a sphere having radius 1 and center $(0, 0, 0)$.

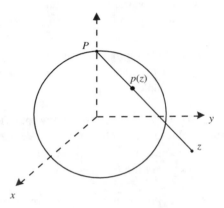

Figure 1.

We associate to each point z in the complex plane the point $p(z)$ of intersection of the line through z and the "north pole" P with the sphere.

If we equip Euclidean 3-space with coordinates as indicated in the figure, then the map p is given explicitly by

$$p(x, y) = \left(\frac{2x}{1 + x^2 + y^2}, \frac{2y}{1 + x^2 + y^2}, 1 - \frac{2}{1 + x^2 + y^2} \right).$$

It is an elegant exercise in differential calculus to check that the metric

$$\sigma(z) = \frac{2}{1 + |z|^2}$$

on $\mathbb{C}$ has the following property: if z, $w \in \mathbb{C}$, then $d_\sigma(z, w)$ is precisely the distance (along a great circular arc on the surface of the sphere) from $p(z)$ to $p(w)$. In fact a calculation shows that this distance is

$$d_\sigma(z, w) = 2 \arctan\left(\left|\frac{z - w}{1 + \overline{w}z}\right|\right). \qquad (*)$$

Differentiation of the formula $(*)$ along a line in the plane gives rise to the definition of σ. (Be reassured that σ can also be thought of as the pullback of the surface Euclidean metric τ on the sphere by the map p. However, a precise formulation of this assertion would take us far afield and we omit it.) As an exercise, the reader may wish to derive the formula for $d_\sigma(z, w)$ by imitating our derivation of the formula for Poincaré distance (Proposition 2 of Section 1.4): first calculate the special case $z = 0$, $w = r + 0i$, then use the invariance of the metric under conformal maps of the Riemann sphere $\widehat{\mathbb{C}}$ (i.e., under suitable linear fractional transformations).

Now suppose that in our study of normal families we study not complex-valued holomorphic functions but meromorphic functions. A meromorphic function is nothing other than a holomorphic function taking values in the Riemann sphere $\hat{\mathbb{C}}$. In order to make this last point clear, let $m(z)$ be a meromorphic function on a domain U with a pole at $P \in U$. Set $I(z) = 1/z$. Then $I \circ m$ is holomorphic in the usual sense in a neighborhood of P. This is precisely the definition of the property "holomorphic" for a $\hat{\mathbb{C}}$-valued function. Note that if the pole at P is of order k then $I \circ m$ has a zero at P of order k, and conversely.

With this increased generality we can now give a more elegant definition of "normal family." This new definition captures the spirit of the first one, but is strictly more general.

Definition 3$'$. Let $U \subseteq \mathbb{C}$ be a domain. Let $\mathcal{F}$ be a family of functions $f : U \to \widehat{\mathbb{C}}$. Then $\mathcal{F}$ is called a *normal family* if every sequence of elements in $\mathcal{F}$ has a subsequence that converges uniformly on compact sets to a function from U to $\widehat{\mathbb{C}}$. [Note that, in order to make sense of this condition, one must use the provided metric on the range, which is the Riemann sphere.]

Exercises. The point of our new definition, and you should prove this as an exercise, is that a compactly divergent sequence (according to Definition 2 above) is actually normally convergent to the point $\infty \in \hat{\mathbb{C}}$. You should note that if we specialize down to holomorphic functions, then Hurwitz's theorem guarantees that any family which is normal according to Definition 3′ is also normal according to Definition 3.

As a second exercise, discuss Example 1 in light of this new definition. ∎

Now we can give Marty's characterization of normal families in terms of the spherical metric.

Theorem 5. *Let U be a planar domain and let $\mathcal{F}$ be a family of $\hat{\mathbb{C}}$-valued holomorphic functions on U (equivalently, meromorphic functions). Let τ be the standard induced Euclidean metric on the surface of the Riemann sphere. The family $\mathcal{F}$ is normal if and only if the family of pullback metrics*

$$\{f^*\tau : f \in \mathcal{F}\}$$

is uniformly bounded on compact subsets of U.

An equivalent, and more concrete, way of saying this is that $\mathcal{F}$ is a normal family if and only if, for each compact $K \subseteq U$, there is a constant M_K such that for all $z \in K$ and all $f \in \mathcal{F}$ we have

$$\frac{2|f'(z)|}{1 + |f(z)|^2} \leq M_K.$$

Proof of the Theorem. The equivalence of the two statements follows immediately from the discussion of the spherical metric and the definition of pullback metric. We now prove the second formulation of the theorem.

First we fix a compact set K and assume that the inequality

$$\frac{2|f'(z)|}{1 + |f(z)|^2} \leq M_K, \qquad \forall f \in \mathcal{F}, z \in K, \tag{$*$}$$

holds. Without loss of generality we may assume that K is the closure of a connected open set V. If $\gamma : [a, b] \to V$ is a continuously differentiable curve then the spherical length of $f \circ \gamma$ is

$$\int_a^b \| (f \circ \gamma)'(t) \|_{\tau, f \circ \gamma(t)} \, dt.$$

It follows from $(*)$ that this quantity does not exceed

$$\int_a^b M_K \cdot |\gamma'(t)| \, dt.$$

But this last is just M_K times the Euclidean length of γ. Notice that this estimate is independent of $f \in \mathcal{F}$. By our definition of distance, it follows that $\mathcal{F}$ is an equicontinuous family of functions when thought of as mapping K equipped with the Euclidean metric to $\hat{\mathbb{C}}$ equipped with the spherical metric. A metric-space-valued version of the Ascoli/Arzelà theorem (really just the same as the one we discussed in Section 0.3) now yields immediately that $\mathcal{F}$ is a normal family.

Now we treat the converse direction. Suppose that $\mathcal{F}$ is a normal family. Define the expression (the *spherical derivative*)

$$f^{\#}(z) = \frac{2|f'(z)|}{1 + |f(z)|^2}.$$

Fix a compact set $K \subseteq U$. Seeking a contradiction, we suppose that the functions $\{f^{\#} : f \in \mathcal{F}\}$ are not uniformly bounded on K. Then there is a sequence $f_j \in \mathcal{F}$ such that the maximum of $(f_j)^{\#}$ on K tends to infinity with j. By the normal family hypothesis, we may suppose that each point $P \in U$ has a neighborhood N_P on which $\{f_j\}$ converges normally to a holomorphic function f which takes values in $\hat{\mathbb{C}}$. But then $(f_j)^{\#}$ converges normally on each N_P (it is convenient here to note that $(f_j)^{\#} = (1/f_j)^{\#}$). Since K is compact, we may cover K with finitely many of the sets N_P. It follows that $(f_j)^{\#}$ is bounded on K, contradicting our assumption. $\blacksquare$

Remark. If g is a meromorphic function with a pole at P then the spherical derivative $g^{\#}(P)$ is defined to be $\lim_{z \to P} g^{\#}(z)$. It is an exercise for the reader to confirm that this limit will always exist. ■

The thrust of Marty's theorem (and its *proof*) is that the condition of being a normal family is really an equicontinuity condition on the family, *when measured in the correct metric*. That metric becomes plain when the functions are viewed as taking values in $\hat{\mathbb{C}}$.

In the next section we will derive two striking consequences of the work done here.

4. A Generalization of Montel's Theorem and the Great Picard Theorem

Now we reap the reward of our labors. We obtain simple proofs of two of the great results of classical function theory, theorems whose traditional proofs (involving the elliptic modular function, for instance) are quite elaborate. Our main tools here will be the generalized Schwarz lemma and the work in Section 3. We begin with a more general version (also due to Paul Montel) of the Montel theorem which we discussed earlier:

Theorem 1. *Let U be a domain in the complex plane. Let P, Q, and R be distinct elements of $\hat{\mathbb{C}}$. Suppose that $\mathcal{F}$ is a family of holomorphic functions taking values in $\hat{\mathbb{C}} \setminus \{P, Q, R\}$. Then $\mathcal{F}$ is a normal family in the sense of Definition 3' of Section 3.*

Proof. We may apply a linear fractional transformation to arrange for $P = 0$, $R = \infty$, and $Q = 1$. Thus we need to show that a family of holomorphic functions taking values in $\mathbb{C}_{0,1} \equiv \mathbb{C} \setminus \{0, 1\}$ is normal.

Of course it is enough to show that $\mathcal{F}$ is normal on any disc $D(z_0, \alpha) = \{z : |z - z_0| < \alpha\} \subseteq U$. We may assume that $z_0 = 0$.

Let μ be the metric of strictly negative curvature on $\mathbb{C}_{0,1}$ constructed in Theorem 3 of Section 2. After multiplying μ by a positive constant, we may assume that the curvature is bounded from above by -4. By Theorem 4 of Section 1 (the generalized Schwarz lemma), with $A = 4$ and $B = 4$, we have for any $f \in \mathcal{F}$ that

$$f^*\mu(z) \leq \rho_\alpha^A(z), \qquad \forall z \in D(0, \alpha).$$

Now let us compare the spherical metric $\sigma(z)$ restricted to $\mathbb{C}_{0,1}$ with the metric μ. Notice that $\sigma(z)/\mu(z)$ tends to zero as z tends to either 0, 1, or ∞. It follows that there is a constant $M > 0$ such that

$$\sigma(z) \leq M \cdot \mu(z).$$

Therefore we have

$$f^\# \equiv f^*\sigma \leq M \cdot f^*\mu \leq M \cdot \rho_\alpha^A \qquad \text{on } D(0, \alpha).$$

The constant M is independent of $f \in \mathcal{F}$. Thus $f^\#$ is bounded on each compact subset of $D(0, \alpha)$, with bound independent of $f \in \mathcal{F}$. Now Marty's theorem implies that $\mathcal{F}$ is a normal family. ∎

We isolate the case $R = \infty$ of the theorem as a corollary.

Corollary 1.1. *If $\mathcal{F}$ is a family of complex-valued holomorphic functions on U all of which omit the same two complex values from their images, then $\mathcal{F}$ is a normal family.*

We can now use this generalized Montel theorem to obtain the refinement of Picard's theorem which was promised at the end of Section 2. This result is known as Picard's great theorem.

Theorem 2. (Picard). *Let $U = D'(0, \alpha) \equiv D(0, \alpha) \setminus \{0\}$ be a punctured disc and let f be a holomorphic function on U which has an essential singularity at 0. Then f restricted to any deleted neighborhood of 0 omits at most one complex value from its image.*

Proof. We prove the contrapositive. After rescaling, we may assume that we have an f on $D' \equiv D(0, 1) \setminus 0$ such that the image of f omits the values 0 and 1. We will prove that 0 is therefore either a pole or a removable singularity for f, resulting in a contradiction.

Define $f_n(z) = f(z/n), 0 < |z| < 1$, and let $\mathcal{F} = \{f_n\}$. Since all elements of $\mathcal{F}$ take values in $\mathbb{C}_{0,1}$, the family $\mathcal{F}$ is normal. Because the family $\mathcal{F}$ consists of holomorphic functions, we conclude that there is a subsequence f_{n_k} which either converges normally or diverges compactly.

In the first instance, the subfamily $\{f_{n_k}\}$ is bounded on compact subsets of D'. In particular, it is bounded by some constant M on the set $\{z : |z| = 1/2\}$. But this says that f itself is bounded by M on each of the circles $\{z : |z| = 1/(2n_k)\}$. By the maximum principle, f is bounded by M on $\{z : 0 < |z| < 1/2(n_1)\}$. So 0 is a removable singularity for f.

In the second case, a similar argument applied to $1/f$ shows that f tends to ∞ as $z \to 0$. Therefore, 0 is a pole for f.

We have shown that if f omits two values in a deleted neighborhood of 0, then f has either a removable singularity or a pole at 0. This completes the proof of the contrapositive of the theorem. ∎

CHAPTER 3

Some New Invariant Metrics

0. Introductory Remarks

Refer to Section 0.3 for the statement and sketch of the proof of the Riemann mapping theorem. The Riemann mapping function is the solution to a certain extremal problem: to find a map of the given domain U into the disc D which is one-to-one, maps a given point P to 0, and has *derivative of greatest possible modulus* λ_P at P. The existence of the extremal function, which also turns out to be one-to-one, is established by normal families arguments; the fact that the extremal function is onto is established by an extra argument which is in fact *the only step of the proof where the topological hypotheses about U are used.*

The point of the present discussion is to observe that the scheme we just described can be applied *even* if U is not topologically equivalent to the disc. Constantin Carathéodory's brilliant insight was that the quantity λ_P can be used to construct a metric, now called the Carathéodory metric. We maximize the derivative at P of maps ϕ of U into D such that $\phi(P) = 0$ *but we no longer require the maps to be injective.* Of course the proof of the Riemann mapping theorem will break down at the stage where we attempt to show that the limit map is surjective; we will also be unable to prove that it is injective. All other steps, *including the existence of the extremal function,* are correct and give rise to a metric (as in Section 3.1 following) on the domain.

A dual construction, using maps $\phi : D \to U$, results in a metric called the Kobayashi or Kobayashi/Royden metric. Both the Carathéodory and the Kobayashi/Royden metrics are motivated by the extremal problem which arises from the Riemann mapping theorem, and are of interest because they endow virtually any domain with an invariant metric. Moreover, these metrics turn out to be invariant under conformal mappings (more generally, they are distance-decreasing under arbitrary holomorphic mappings). But the richest benefit is that which can be obtained from studying the interaction of the two metrics.

In this chapter we learn about the Carathéodory and Kobayashi metrics and the insights they lend to the theory of holomorphic mappings. We will also begin to explore what can be learned by comparing the two metrics.

Note that in Chapter 1 we defined a "metric" to be a twice continuously differentiable function with certain other special properties. This requirement was imposed in part so that we could study curvature. For the Carathéodory and Kobayashi metrics, we shall not examine curvature. So we may suspend the smoothness hypothesis.

1.   The Carathéodory Metric

Fix a domain $U \subseteq \mathbb{C}$. Recall that $D \subseteq \mathbb{C}$ is the unit disc.

Definition 1. If $P \in U$, define

$$(D, U)_P = \{\text{holomorphic functions } f \text{ from } U \text{ to } D$$
$$\text{such that } f(P) = 0\}.$$

The *Carathéodory metric* for U at P is defined to be

$$F_C^U (P) = \sup\{|\phi'(P)| : \phi \in (D, U)_P\}.$$

Remark. As predicted in Section 0, the quantity F_C^U measures, for each P, the extreme value posited in the proof of the Riemann mapping theorem. In that proof, it was necessary only to know that the extreme value existed and was finite. Now we shall gain extra information by comparing this value to other quantities. ∎

Clearly $F_C^U(P) \geq 0$. (Moreover the Cauchy estimates imply that $F_C^U(P) < \infty$.) Is $F_C^U(P) > 0$ for all P? If U is bounded, then the answer is "yes." For then

$$U \subseteq D(0, R) = \{z \in \mathbb{C} : |z| < R\}$$

for some $R > 0$. But then the map

$$\phi : \zeta \longmapsto \frac{\zeta - P}{2R}$$

satisfies $\phi \in (D, U)_P$. Thus

$$F_C^U(P) \geq |\phi'(P)| = \frac{1}{2R} > 0.$$

However, if U is unbounded, then F_C^U may degenerate. If $U = \mathbb{C}$, for instance, then any $f \in (D, U)_P$ is constant, hence $F_C^U \equiv 0$. The same holds for U equaling $\mathbb{C}$ less a discrete set of points (by the Riemann removable singularities theorem). Which domains have nondegenerate (i.e., non-vanishing) Carathéodory metric (i.e., Carathéodory metric which actually gives a distance)? Analytic capacity is a device for answering this question (see [GAR]), but we cannot pursue that topic here.

The primary interest, at least right at the moment, of the Carathéodory metric is that it generalizes Proposition 1 in Section 1.5 to arbitrary domains.

Proposition 2. *Let U_1, U_2 be domains in $\mathbb{C}$. Let ρ_j be the Carathéodory metric on U_j. If $h : U_1 \to U_2$ is holomorphic, then h is distance-decreasing from (U_1, ρ_1) to (U_2, ρ_2). In other words,*

$$h^* \rho_2(z) \leq \rho_1(z), \qquad \forall z \in U_1.$$

Corollary 2.1. *If $\gamma : [0, 1] \to U_1$ is a piecewise continuously differentiable curve, then*

$$\ell_{\rho_2}(h_*\gamma) \le \ell_{\rho_1}(\gamma).$$

Remark. Observe that the Carathéodory metric is not manifestly integrable on curves. But if one thinks about the fact that it is constructed as the supremum of continuous functions, then it follows that the metric is semicontinuous (see [KR1]). So it is indeed integrable on curves (to wit, a semi-continuous function is the monotone limit of continuous functions). ∎

Corollary 2.2. *If P_1, $P_2 \in U_1$ then*

$$d_{\rho_2}(h(P_1), h(P_2)) \le d_{\rho_1}(P_1, P_2).$$

Corollary 2.3. *If h is conformal, then h is an isometry of (U_1, ρ_1) to (U_2, ρ_2).*

The corollaries are proved just as they were proved for the Poincaré metric in Chapter 1. We concentrate on proving the Proposition.

Proof of Proposition 2. Fix $P \in U_1$ and set $Q = h(P)$. Notice that if $\phi \in (D, U_2)_Q$, then $\phi \circ h \in (D, U_1)_P$. Thus

$$F_C^{U_1}(P) \ge |(\phi \circ h)'(P)|$$
$$= |\phi'(Q)| \cdot |h'(P)|.$$

Taking the supremum over all $\phi \in (D, U_2)_Q$ yields

$$F_C^{U_1}(P) \ge F_C^{U_2}(Q) \cdot |h'(P)|$$

or

$$\rho_1(P) \ge h^*\rho_2(P). \qquad ∎$$

Next we calculate the Carathéodory metric for the disc.

Proposition 3. *The Carathéodory metric on the disc coincides with the Poincaré metric.*

Proof. First we calculate the metric at the origin. If $\phi \in (D, D)_0$ then, by the Schwarz lemma, $|\phi'(0)| \leq 1$. Therefore

$$F_C^D(0) \leq 1.$$

But the map

$$\phi(\zeta) = \zeta$$

satisfies $\phi \in (D, D)_0$ and $\phi'(0) = 1$. Thus

$$F_C^D(0) = 1.$$

Because every conformal map of the disc is an isometry of the Carathéodory metric, Proposition 6 of Section 1.4 now implies that the Carathéodory metric and the Poincaré metric are equal. ∎

Given what we have learned so far, we might suspect that any isometry of the Carathéodory metric must be a conformal map. This is indeed true—in fact a much stronger assertion holds—provided the domains in question have nondegenerate (i.e., non-vanishing) Carathéodory metrics. We postpone this topic until after we have introduced the Kobayashi metric.

2. The Kobayashi Metric

Fix a domain $U \subseteq \mathbb{C}$.

Definition 1. If $P \in U$, define

$$(U, D)^P = \{\text{holomorphic functions } f \text{ from } D$$
$$\text{to } U \text{ such that } f(0) = P\}.$$

The *Kobayashi* (or Kobayashi/Royden) *metric* for U at P is defined to be

$$F_K^U(P) = \inf\left\{\frac{1}{|\phi'(0)|} : \phi \in (U, D)^P\right\}.$$

Remark. As we remarked in the last section about the Carathéodory metric, the Kobayashi metric quantifies a certain extremal problem and will be useful for comparison purposes. The particular form of the definition of the Kobayashi metric will turn out to facilitate comparisons with the Carathéodory metric; in particular, the two metrics will turn out to interact nicely with the classical Schwarz lemma.   ∎

Clearly $F_K^U(P) \geq 0$. In order to learn more about F_K^U, we now compare it with the Carathéodory metric.

Proposition 2. *For all $P \in U$,*

$$F_C^U(P) \leq F_K^U(P).$$

Proof. Let $\phi \in (D, U)_P$ and $\psi \in (U, D)^P$. Then $\phi \circ \psi : D \to D$ and $\phi \circ \psi(0) = 0$. It follows from Schwarz's lemma that

$$|(\phi \circ \psi)'(0)| \leq 1$$

or

$$|\phi'(P)| \leq \frac{1}{|\psi'(0)|}.$$

Taking the supremum over all ϕ gives

$$F_C^U(P) \leq \frac{1}{|\psi'(0)|}.$$

Now taking the infimum over all ψ yields

$$F_C^U(P) \leq F_K^U(P).$$ ∎

An immediate consequence of this last result is that if U is bounded, then F_K^U is nondegenerate, i.e., non-vanishing (for then F_C^U is). On the other hand, if $U = \mathbb{C}$, then $F_K^U \equiv 0$; for given $P \in \mathbb{C}$ the function

$$\phi_R(\zeta) = P + R\zeta$$

satisfies $\phi_R \in (U, D)_P$, any $R > 0$. Thus

$$F_K^U(P) \leq \frac{1}{|\phi_R'(0)|} = \frac{1}{R};$$

letting $R \to \infty$, we have $F_K^U(P) = 0$.

In analogy with Proposition 2 of the last section we now have:

Proposition 3. *Let U_1, U_2 be domains equipped with the Kobayashi metric (denoted, respectively, by ρ_1 and ρ_2). If $h : U_1 \to U_2$ is holomorphic, then h is distance-decreasing from (U_1, ρ_1) to (U_2, ρ_2). That is,*

$$h^*\rho_2(z) \leq \rho_1(z), \qquad \forall z \in U_1.$$

Corollary 3.1. *If $\gamma : [0, 1] \to U_1$ is a piecewise continuously differentiable curve, then*

$$\ell_{\rho_2}(h_*\gamma) \leq \ell_{\rho_1}(\gamma).$$

Remark. Just as for the Carathéodory metric, the Kobayashi metric will be integrable on curves just because it is semicontinuous. Again see [KR1]. ∎

Corollary 3.2. *If $P_1, P_2 \in U_1$, then*

$$d_{\rho_2}(h(P_1), h(P_2)) \leq d_{\rho_1}(P_1, P_2).$$

Corollary 3.3. *If h is conformal, then h is an isometry of (U_1, ρ_1) to (U_2, ρ_2).*

Proof of Proposition 3. Fix $P \in U_1$ and put $Q = h(P)$. Choose $\phi \in (U_1, D)^P$. Then $h \circ \phi \in (U_2, D)^Q$. We have

$$F_K^{U_2}(Q) \leq \frac{1}{|(h \circ \phi)'(0)|}$$

$$= \frac{1}{|h'(P)||\phi'(0)|}.$$

Taking the infimum over all $\phi \in (U_1, D)^P$ gives

$$F_K^{U_2}(Q) \leq \frac{1}{|h'(P)|} \cdot F_K^{U_1}(P)$$

or

$$(h^* F_K^{U_2})(P) \leq F_K^{U_1}(P). \qquad \blacksquare$$

Using Proposition 3, we may give another example of domains $U \subseteq \mathbb{C}$ with degenerate Kobayashi metric. Let $\mathbb{C}_0 = \mathbb{C} \setminus \{0\}$. Then

$$h : \mathbb{C} \to \mathbb{C}_0$$

$$\zeta \to e^\zeta$$

is holomorphic. The map h will be distance-decreasing in the Kobayashi metric. Since F_K is identically 0 on $\mathbb{C}$ (because there are arbitrarily large analytic discs in $\mathbb{C}$ centered at any point), it follows that F_K is identically zero on $\mathbb{C}_0$.

As an exercise, prove that we cannot continue this line of reasoning to see that F_K^U is degenerate when $U = \mathbb{C} \setminus \{0, 1\}$. In fact, we shall later see that F_K^U is nondegenerate on $U \equiv \mathbb{C} \setminus \{P_1, \dots, P_k\}$ provided the P_j's are distinct and $k \geq 2$ (notice that this situation is in contrast to that for the Carathéodory metric as discussed in Section 1). The matter will be addressed in Section 5; there we will see it connected in an elegant way with normal families—recall the special role of $\mathbb{C}_{0,1}$ in that theory. Meanwhile, we turn our attention to more elementary questions regarding the Kobayashi metric.

Proposition 4. *For* $U = D \subseteq \mathbb{C}$, *the Kobayashi metric equals the Poincaré metric.*

Proof. For $P \in D$ we have

$$F_K^D(P) \geq F_C^D(P) = \rho(P),$$

where ρ is the Poincaré metric on the disc. For an inequality in the opposite direction, first take $P = 0$. Let $\phi \in (D, D)^0$ be given by

$$\phi(\zeta) = \zeta.$$

Then

$$F_K^D(0) \leq \frac{1}{|\phi'(0)|} = 1 = \rho(0).$$

It follows that

$$F_K^D(0) = 1 = \rho(0).$$

By Corollary 3.3 above and Proposition 6 of Section 1.4, we conclude that

$$F_K^D(P) = \rho(P), \qquad \forall P \in D. \qquad \blacksquare$$

The first major result of this chapter is a characterization of the disc in terms of the Carathéodory and Kobayashi metrics (at this point have another look at Section 0 to see why this result is a metric version of the Riemann mapping theorem):

Theorem 5. *Let* $U \subseteq \mathbb{C}$ *be a domain. The domain* U *is conformally equivalent to the disc if and only if there is a point* $P \in U$ *such that*

$$F_C^U(P) = F_K^U(P) \neq 0.$$

Proof. If U is conformally equivalent to the disc then let

$$h : U \to D$$

be a conformal map. For any point $P \in U$ we have that

$$F_C^U(P) = (h^* F_C^D)(P)$$

which, by Proposition 3 of Section 1,

$$= (h^* \rho)(P);$$

but by Proposition 4 this

$$= (h^* F_K^D)(P).$$

By Corollary 3 to Proposition 3 we conclude that this last equals $F_K^U(P)$, proving the easy half of the theorem.

For the converse, choose $\phi_j \in (D, U)_P$ such that

$$|\phi_j'(P)| \to F_C^U(P)$$

and choose $\psi_j \in (U, D)^P$ such that

$$\frac{1}{|\psi_j'(0)|} \to F_K^U(P).$$

Since $\{\phi_j\}$ is uniformly bounded above by 1, we may extract a subsequence $\{\phi_{j_k}\}$ converging to a normal limit ϕ_0. Consider

$$h_{j_k} = \phi_{j_k} \circ \psi_{j_k}.$$

Passing to another subsequence, which we denote by h_{j_ℓ}, we may suppose that h_{j_ℓ} converges normally to a limit h_0. Notice that $h_0(0) = 0$ so that h_0 is not a unimodular constant; therefore h_0 maps D to D. After renumbering, we call this last sequence

$$h_\ell = \phi_\ell \circ \psi_\ell.$$

Recall that when a sequence of holomorphic functions converges normally, then so does the sequence of its derivatives (see Corollary 5.1 in

Section 0.1). It follows that

$$|h_0'(0)| = \lim_{\ell \to \infty} |(\phi_\ell \circ \psi_\ell)'(0)|$$

$$= \lim_{\ell \to \infty} |\phi_\ell'(P)| \cdot |\psi_\ell'(0)|$$

$$= F_C^U(P) \cdot \frac{1}{F_K^U(P)}$$

$$= 1.$$

By Schwarz's lemma, $h_0(\zeta) = \mu \cdot \zeta$ for some unimodular constant μ. Thus we have

$$\mu \cdot \zeta = h_0(\zeta) = \lim_{\ell \to \infty} (\phi_\ell \circ \psi_\ell(\zeta)). \tag{$*$}$$

After composing the functions ϕ_ℓ with a rotation, we may assume that $\mu = 1$.

Now $\mathbb{C} \setminus U$ must contain at least two points (else F_K^U would be identically 0, contradicting our hypothesis). Hence, by Theorem 1 of Section 2.4, $\{\psi_\ell\}$ forms a normal family. Say that a subsequence ψ_{ℓ_m} converges to some ψ_0. After renumbering, we rewrite $(*)$ as

$$\zeta = h_0(\zeta)$$

$$= \lim_{\ell \to \infty} (\phi_\ell \circ \psi_\ell(\zeta)) \tag{$**$}$$

$$= \phi_0 \circ \psi_0(\zeta).$$

Since h_0 is surjective, we conclude that ϕ_0 is surjective.

Because of $(**)$, the function ψ_0 is a nonconstant holomorphic function; hence its image is open. We claim that the image is also closed (in the relative topology of U). To see this, let $\psi_0(\zeta_j)$ be elements of that image converging to a limit point $q \in U$. Since ϕ_0 is continuous, it follows that $\phi_0(\psi_0(\zeta_j))$ converges to a limit point r. But then $(**)$ tells us that $\zeta_j \to r$. It follows from the continuity of ψ_0 that $\psi_0(r) = q$, so q is in the image of ψ_0. We thus have that the image of ψ_0 is both open and closed, and it is nonempty. Since U is connected, it follows that the image of ψ_0 equals U, hence ψ_0 is surjective. Thus, since h_0

is injective, so must ϕ_0 be injective. We conclude that ϕ_0 is the desired conformal map of U to D.   ∎

Notice that the conclusion of the theorem is false if we allow the metrics to degenerate. For example, if $U = \mathbb{C} \setminus 0$, then $F_C^U = F_K^U \equiv 0$, yet U is definitely not conformally equivalent to the disc.

As anticipated in the last section, we now prove that the only isometries of the Carathéodory or Kobayashi metrics which fix a point are conformal maps. In fact we shall prove something much stronger.

Theorem 6. *Let $U \subseteq \mathbb{C}$ be a domain on which the Kobayashi metric is nondegenerate (i.e., gives a genuine distance function) and fix $P \in U$. Suppose that*

$$f : U \longrightarrow U$$

is a holomorphic function such that $f(P) = P$. Assume that f is an isometry of either the Carathéodory or the Kobayashi metric at P; that is, suppose that either

$$f^* F_C^U(P) = F_C^U(P)$$

or

$$f^* F_K^U(P) = F_K^U(P)$$

and that the metric does not vanish at P. Then f is a conformal map of U onto U.

Remark. This theorem is a remarkable rigidity statement: the global condition that f map U to U, together with the differential condition at P, forces f to be one-to-one and onto.   ∎

Proof of the Theorem. Since U has nondegenerate Kobayashi metric, $\mathbb{C} \setminus U$ must contain at least two points. This implies that any family of holomorphic functions $\{g_\alpha\}$ taking values in U is normal (by Theorem 1 of Section 2.4). We will use this observation repeatedly.

Now the hypothesis that

$$f^* F_C^U(P) = F_C^U(P) \quad \text{or} \quad f^* F_K^U(P) = F_K^U(P)$$

implies that

$$|f'(P)| = 1.$$

Define

$$f^1 = f$$
$$f^2 = f \circ f$$
$$\cdots$$
$$f^j = f^{j-1} \circ f, \qquad j \geq 2.$$

Then $\{f^j\}$ is a normal family. Since the numbers $(f^j)'(P)$ all have unit modulus (note that $(f^j)'(P) = [f'(P)]^j$), there is a subsequence $\{f^{j_\ell}\}$ such that $(f^{j_\ell})'(P) \to 1$. (Exercise: if $f'(P)$ has argument which is a rational multiple of π, then the assertion is clear; if it is an irrational multiple of π, then the set of $(f^j)'(P)$ forms a dense subset of the circle.) Passing to another subsequence, which we still denote by $\{f^{j_\ell}\}$, we may also suppose that f^{j_ℓ} converges normally to a holomorphic function $\tilde{f}$. It follows that $\tilde{f}'(P) = 1$. We claim in fact that $\tilde{f}(z) \equiv z$.

To prove this assertion, suppose not. Assume for simplicity that $P = 0$. Then, for z near 0,

$$\tilde{f}(z) = 0 + z + \text{ (higher order terms)}.$$

Let $a_m z^m$, $m \geq 2$, be the first nonvanishing power of z which appears after z (since we are assuming that $\tilde{f}$ is not the function z, there must be one). Notice that

$$\tilde{f}^2 \equiv \tilde{f} \circ \tilde{f} = z + 2a_m z^m + \cdots$$
$$\tilde{f}^3 \equiv \tilde{f} \circ \tilde{f} \circ \tilde{f} = z + 3a_m z^m + \cdots$$
$$\cdots \qquad\qquad\qquad\qquad (*)$$
$$\tilde{f}^k \equiv \tilde{f} \circ \tilde{f} \circ \cdots \circ \tilde{f} = z + k a_m z^m + \cdots$$

On the other hand, $\{\tilde{f}^k\}$ forms a normal family, so there is a subsequence $\tilde{f}^{k_q}$ converging normally on U. Hence

$$\left(\frac{\partial}{\partial z}\right)^m \tilde{f}^{k_q}(P)$$

converges. But (*) clearly shows that

$$\left(\frac{\partial}{\partial z}\right)^m \tilde{f}^k(P) = m! \cdot k \cdot a_m$$

blows up with k. This contradiction can only be resolved if $a_m = 0$. Therefore $\tilde{f}(z) \equiv z$.

Now we have that $f^{j_\ell}(z) \to z$ normally. We claim that this implies that f is a conformal map. Indeed, consider the family $\{f^{j_\ell - 1}\}$. It is normal, so there is a subsequence, call it $\{f^{j_r - 1}\}$, which converges normally to a limit g. Notice that g is not constant since $g'(P)$ is not zero. Then

$$f \circ g(z) = f \circ \lim_{r \to \infty} f^{j_r - 1}(z)$$

$$= \lim_{r \to \infty} f^{j_r}(z)$$

$$\equiv z.$$

Similarly,

$$g \circ f(z) \equiv z.$$

Thus f is one-to-one and onto. It is certainly holomorphic, so it is a conformal map as claimed. ∎

It is instructive to view Theorem 6 as a generalization of the uniqueness part of the Schwarz lemma. For simplicity take U to be a bounded domain. Then U is contained in some disc, and a disc has nondegenerate Kobayashi metric. Therefore, by the distance decreasing property of the Kobayashi metric applied to the inclusion map, the domain U also has nondegenerate Kobayashi metric. Suppose that f

is a holomorphic function mapping U to U that fixes a point P in U. The distance-decreasing property of the Kobayashi metric tells us that $|f'(P)| \le 1$. Theorem 6 now says that equality obtains if and only if f is one-to-one and onto.

3. Completeness of the Carathéodory and Kobayashi Metrics

In this section we prove that if U is a domain with a reasonably nice boundary then it is complete, as a metric space, when equipped with either the Carathéodory or Kobayashi metric. A warning is in order here: this is simultaneously the richest and the most difficult section of the book; for it uses both the metric language and tricky constructions from analysis. But the section is self-contained, and provides an introduction to a lot of important and beautiful material. Moreover, the payoff for the work, in both this section and the next, is more than adequate compensation for the effort expended. As a bonus, we will give a nice application of the ideas (in the context of Bergman geometry) in Section 4.6.

We begin by noting that, on compact sets of a bounded domain, all of our metrics are comparable:

Proposition 1. *Let* $\Omega \subseteq \mathbb{C}$ *be a bounded domain. Let* $L \subseteq \Omega$ *be a compact subset. Let* ρ_C, ρ_K, *and* ρ_E *be the Carathéodory, Kobayashi, and Euclidean metrics on* Ω *respectively. Then there are constants* C_1, C_2, C_3, C_4 *such that, for* $z \in L$,

$$C_1 \le \frac{\rho_C(z)}{\rho_E(z)} \le C_2$$

and

$$C_3 \le \frac{\rho_K(z)}{\rho_E(z)} \le C_4.$$

[We stress that these four constants certainly depend both on L *and on* Ω.*]*

Proof. We will prove the second set of inequalities. The proof of the first inequalities is similar.

Now let P be a point in Ω. Let r be a positive number that is less than one third of the Euclidean distance of P to the boundary. Let $\overline{D}(P, r) \subseteq \overline{D}(P, 2r)$ be closed discs in Ω. If $z \in D(P, r)$ then

$$\rho_K^{D(P,2r)}(z) \geq \rho_K^{\Omega}(z).$$

Here we use a superscript on ρ to indicate with respect to which domain the metric is being computed.

On the other hand,

$$\rho_K^{D(P,2r)}(z) = \frac{1}{2r} \cdot \rho_K^{D(0,1)}((z - P)/[2r]).$$

And of course we know the Kobayashi metric on the unit disc explicitly. It equals the Poincaré metric. Hence $\rho_K^{D(0,1)}((z - P)/[2r]) \leq 4/3$. Putting our estimates together, we find that

$$\rho_K^{\Omega}(z) \leq \frac{4}{3} \cdot \frac{1}{2r}.$$

This is one half of our estimate.

For the other half, notice that since Ω is bounded it is contained in some large disc open $D(0, R)$. Then, for $z \in D(P, r)$,

$$\rho_K^{\Omega}(z) \geq \rho_K^{D(0,R)}(z).$$

Now, by conformal mapping, this last is equal to $[1/R]\rho_K^{D(0,1)}(z/[R])$. Thus it is at least equal to $1/R$.

In summary, we have proved for points $z \in D(P, r)$ that

$$\frac{4}{6r} \geq \rho_K^{\Omega}(z) \geq \frac{1}{R}.$$

The compact set $L \subset \Omega$ can be covered by finitely many discs of the form $D(P, r)$. So, in the end, we obtain a uniform estimate

$$C_3 \leq \rho_K^{\Omega}(z) \leq C_4$$

for all $z \in L$. But the Euclidean metric is constantly equal to 1. So this last line just says

$$C_3 \rho_E(z) \leq \rho_K^\Omega(z) \leq C_4 \rho_E(z),$$

and that is what we wished to prove. ∎

Corollary 1.1. *Let $\Omega \subseteq \mathbb{C}$ be a bounded domain. The topology induced by any of the Carathéodory, Kobayashi, or Euclidean metrics is equivalent to the other two.*

Proof. Since the metrics are comparable, the induced balls are comparable. The balls form a sub-basis for the topology. So the topologies are comparable. ∎

Now we turn to the treatment of completeness of our two new metrics on a reasonable class of domains. It is enough to prove completeness for the Carathéodory metric, since any sequence which is Cauchy in the Kobayashi metric is also Cauchy in the Carathéodory metric (exercise: provide the details using Proposition 2 of Section 2). We begin by defining what we mean by "reasonably nice boundary."

Definition 2. A curve $\gamma : [a, b] \to \mathbb{C}$ is called a *closed, twice continuously differentiable curve* if γ is twice continuously differentiable (with one-sided derivatives as usual at the endpoints), if $\dot{\gamma}$ is never 0, if $\gamma(a) = \gamma(b)$, and if the one-sided derivatives of γ, up to and including order 2, at a equal those at b. We write "γ is C^2 and closed." More generally, we may define γ to be *closed, k times continuously differentiable* (for $k \geq 1$) if γ is k times continuously differentiable, (with one-sided derivatives as usual at the endpoints), if $\dot{\gamma}$ is never 0, if $\gamma(a) = \gamma(b)$, and if the one sided derivatives of γ, up to and including order k, at a equal those at b. We write "γ is C^k and closed."

Definition 3. A domain is said to have C^k *boundary* if the boundary consists of finitely many C^k, closed curves.

In case a function is C^k for every k, or a boundary is C^k for every k, then we say that the function or boundary is C^∞ or infinitely differentiable.

In geometric analysis, an alternative definition of C^k boundary often proves to be convenient. We think of a domain U with C^k (resp. C^k) boundary as one which can be specified by

$$U = \{z \in \mathbb{C} : \rho(z) < 0\},$$

with ρ a C^k function such that $\nabla\rho \neq 0$ on ∂U. We call ρ a *defining function* for U. For example, the disc may be specified in this way as

$$D = \{z \in \mathbb{C} : \rho(z) = |z|^2 - 1 < 0\}.$$

A bounded domain with infinitely many holes will *not* have C^k boundary, $k \geq 1$, since any defining function will fail to be smooth at an accumulation point of the holes. See [KRP2] for a thorough treatment of defining functions.

Example 1. The curve $\gamma(t) = e^{it}$, $0 \leq t \leq 2\pi$, is a closed, twice continuously differentiable curve (indeed, it is C^k for every k, or infinitely differentiable). See Figure 1. Notice that the agreement of the first and second derivatives at the endpoints causes a smooth transition in the figure at the point $1 + i0$ where the two ends meet.

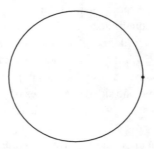

Figure 1.

The curve

$$\mu(t) = (\cos 2t)(\cos t) + i(\cos 2t)(\sin t), \qquad -\frac{\pi}{4} \le t \le \frac{\pi}{4}$$

is closed, but *not* twice continuously differentiable (because the derivatives at $t = -\pi/4$ and $t = \pi/4$ do not agree). The curve is shown in Figure 2. ∎

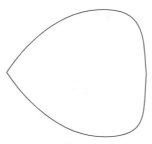

Figure 2.

Figure 3 exhibits a domain with C^k boundary. Note that there are finitely many boundary curves, and each is k times continuously differentiable.

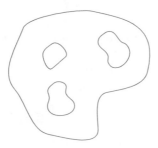

Figure 3.

Observe that, by the usual constructions of multi-variable calculus, a domain with twice continuously differentiable boundary possesses at each boundary point P a well defined unit outward normal ν_P and a well defined unit inward normal ν'_P. See Figure 4.

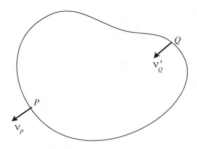

Figure 4.

The functions

$$\partial U \ni P \longrightarrow \nu_P \in \mathbb{R}^2$$

and

$$\partial U \ni P \longrightarrow \nu'_P \in \mathbb{R}^2$$

are then continuously differentiable.

We shall need two geometric/analytic results about domains with C^2 boundary. They are formulated as Propositions 4 and 5.

Proposition 4. *If U is a domain with C^2 boundary, then there is an open neighborhood W of ∂U such that if $z \in U \cap W$, then there is a unique point $P = P(z) \in \partial U$ which is nearest (in the sense of Euclidean distance) to z. In other words,*

$$\inf\{|z - Q| : Q \in \partial U\} = |z - P|.$$

We call W a tubular neighborhood of ∂U (see [MUN] for more on these matters). See Figure 5.

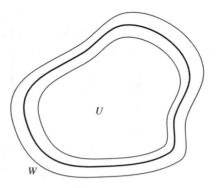

Figure 5.

Proof. Define

$$T : \partial U \times (-1, 1) \longrightarrow \mathbb{C}$$

$$(Q, t) \longmapsto Q + t\nu_Q.$$

Think of T as a map from the two-dimensional real space $\partial U \times (-1, 1)$ to the two-dimensional real space $\mathbb{C} \approx \mathbb{R}^2$. See Figure 6. [In fact the reader may find it convenient to consider this alternative definition of T. Let $\gamma : [0, 1] \to \mathbb{C}$ be a C^2 parametrization of ∂U. Then set

$$T : [0, 1] \times (-1, 1) \longrightarrow \mathbb{C}$$

$$(s, t) \longmapsto \gamma(s) + t\nu_{\gamma(s)}.]$$

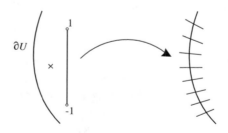

Figure 6.

Fix a point $Q_0 \in \partial U$. Assume without loss of generality that the tangent line to ∂U at Q_0 is horizontal (this may be achieved by rotating the coordinate system). We may also normalize the tangent vector to be of unit length.

Then the Jacobian matrix of T at $(Q_0, 0)$ is

$$\begin{pmatrix} 1 & 0 \\ 0 & 1 \end{pmatrix}$$

which is invertible. By the inverse function theorem (see [RU1] or [MUN] or [KRP3]), there is a neighborhood V_0 of $(Q_0, 0)$ in $\partial U \times (-1, 1)$ on which T is invertible. Then $T(V_0) \equiv W_{Q_0}$ is a neighborhood of Q_0 with the property that if $z \in W_{Q_0}$, then there is a unique point $Q \in \partial U$ and a unique $t \in (-1, 1)$ such that

$$z = Q + t\nu_Q.$$

It follows that Q is (locally) the unique nearest point in ∂U to z and the distance of z to Q is $|t|$. We set

$$W = \bigcup_{Q_0 \in \partial U} W_{Q_0}$$

and we are finished.   ∎

Remark. The hypothesis of C^2 boundary was needed in order to apply the inverse function theorem. Explain why. The analogous result is false as soon as the boundary smoothness is less than C^2 (see [KRP4]).

∎

Proposition 5. *Let $U \subseteq \mathbb{C}$ be a bounded domain with C^2 boundary. Then there is an $r_0 > 0$ such that for each $P \in \partial U$ there is a disc $D(C(P), r_0)$ of radius r_0 which is externally tangent to ∂U at P. There is also a disc $D(C'(P), r_0)$ which is internally tangent to ∂U at P. These discs have the further property that $\overline{D}(C(P), r_0) \cap \partial U = \{P\}$ and $\overline{D}(C'(P), r_0) \cap \partial U = \{P\}$.*

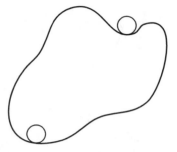

Figure 7.

Remark. Refer to Figure 7 and notice that a disc **d** is internally tangent to ∂U at P if $\mathbf{d} \subseteq U$, $\overline{\mathbf{d}} \ni P$, and $\partial \mathbf{d}$ and ∂U have the same tangent line at P.

A similar remark applies to externally tangent discs.   ∎

Proof of Proposition 5. This follows from basic calculus: To each point $P \in \partial U$ there corresponds a radius of curvature $r(P)$ and a center of curvature $c(P)$ (here we are using the word "curvature" in the classical Euclidean sense). The point $c(P)$ is $r(P)$ units along the principal normal from P.

The corresponding circle is called the circle of curvature. See [THO] for details. If the circle of curvature is *inside* the domain, take $C'(P) = c(P)$ and take $C(P)$ to be the reflected point in ∂U. See Figure 8. If the circle of curvature is *outside* the domain, take $C(P) = c(P)$ and take $C'(P)$ to be a reflected point in ∂U.

Now, referring to [THO], we see that $r(P)$ depends on the second derivative of the boundary curve γ. In particular $r(P)$ is a continuous function of P. We take r_0 to be the (positive) minimum of r. We would like to think that we are finished.

Unfortunately, in spite of our best intentions, we may have a situation as in Figure 9 or 10: the circle of curvature has the right behavior *at P*, but it does not take into account the global behavior of U. As a result, it may be neither completely internal to U nor completely ex-

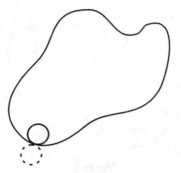

Figure 8.

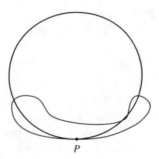

Figure 9.

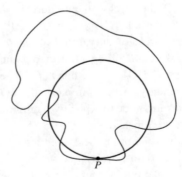

Figure 10.

ternal to U. We therefore make the following adjustment. Let W be a tubular neighborhood of ∂U and choose $\epsilon > 0$ such that if $z \in \mathbb{C}$ has Euclidean distance less than ϵ from ∂U, then z is in W.

Define

$$r^*(P) = \min\{\epsilon/2, r(P)\}.$$

This is a positive, continuous function of $P \in \partial U$. Hence there is a number $r_0 > 0$ such that

$$r^*(P) > r_0 \qquad \forall P \in \partial U.$$

This is the r_0 we seek. We redefine

$$C(P) = P + r_0 \nu_P \quad \text{and} \quad C'(P) = P + r_0 \nu'_P.$$

The definition of tubular neighborhood guarantees that a disc with center $C(P)$ or $C'(P)$ and radius r_0 can intersect ∂U only at the point P. ∎

We return to our discussion of metrics. The main result of this section is the following.

Theorem 6. *If $U \subseteq \mathbb{C}$ is a bounded domain with C^2 boundary, then U is complete in the Carathéodory metric.*

Corollary 6.1. *The domain U is also complete in the Kobayashi metric.*

Proof of the Theorem. Let $z \in U$ be in a tubular neighborhood W of ∂U, as provided by Proposition 4. Let P be the nearest boundary point to z and $D(C(P), r_0)$ the externally tangent disc provided by Proposition 5. The map

$$\mathbf{i}_P : U \longrightarrow D(C(P), r_0)$$

$$\zeta \longmapsto C(P) + \frac{(r_0)^2}{\zeta - C(P)}$$

is holomorphic and inverts U into the disc $D(C(P), r_0)$. The map

$$\mathbf{j}_P : D(C(P), r_0) \longrightarrow D(0, 1)$$
$$\zeta \longmapsto \frac{\zeta - C(P)}{r_0}$$

is holomorphic. To estimate the Carathéodory metric at z, we use the distance-decreasing property of the Carathéodory metric:

$$F_C^U(z) \geq \left((\mathbf{j}_P \circ \mathbf{i}_P)^* F_C^{D(0,1)} \right)(z) \tag{$*$}$$
$$\equiv |(\mathbf{j}_P \circ \mathbf{i}_P)'(z)| F_C^{D(0,1)}(\mathbf{j}_P(\mathbf{i}_P(z))).$$

We now estimate the various terms in this last expression.

First,

$$\left| (\mathbf{j}_P \circ \mathbf{i}_P)'(z) \right| = \left| \frac{1}{r_0} \cdot (\mathbf{i}_P)'(z) \right|$$
$$= \left| \frac{1}{r_0} \cdot \frac{(r_0)^2}{(z - C(P))^2} \right| \tag{$**$}$$
$$= \left| \frac{r_0}{(z - C(P))^2} \right|$$
$$= \left| \frac{r_0}{(\delta + r_0)^2} \right|,$$

where δ is the Euclidean distance of z to P. See Figure 11.

Also

$$\mathbf{j}_P \circ \mathbf{i}_P(z) = \mathbf{j}_P \left(\frac{(r_0)^2}{z - C(P)} + C(P) \right)$$
$$= \frac{r_0}{z - C(P)}$$

hence

$$|\mathbf{j}_P \circ \mathbf{i}_P(z)| = \frac{r_0}{\delta + r_0} = 1 - \frac{\delta}{r_0 + \delta}.$$

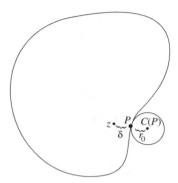

Figure 11.

Recall that, on the unit disc, the Carathéodory metric coincides with the Poincaré metric. It follows from our calculation of the Poincaré metric in Proposition 2 of Section 1.4 that

$$F_C^{D(0,1)}(\mathbf{j}_P \circ \mathbf{i}_P(z)) = \frac{1}{(\delta/(r_0 + \delta))(2 - \delta/(r_0 + \delta))}$$

$$\geq \frac{r_0 + \delta}{2\delta} \qquad (***)$$

$$\geq \frac{r_0}{2} \cdot \frac{1}{\delta}.$$

In summary, using $(*)$, $(**)$, and $(***)$, we have

$$F_C^U(z) \geq \frac{r_0}{(\delta + r_0)^2} \cdot \frac{r_0}{2} \cdot \frac{1}{\delta} \geq C_0 \cdot \frac{1}{\delta},$$

where C_0 is a positive constant which depends only on r_0.

But this is precisely the estimate which enabled us, in Section 1.4, to prove that the disc is complete in the Poincaré metric. We leave it now as an exercise to provide the details which show that the Carathéodory distance from any fixed interior point $P_0 \in U$ to a point with Euclidean distance δ from the boundary has size $C \cdot (1 + |\log 1/\delta|)$ and to conclude that U is complete in the Carathéodory metric. ∎

Exercise. Exploit the internally tangent disc at each boundary point to prove that there is a constant C_1 such that

$$F_C^U(z) \leq C_1 \cdot \frac{1}{\delta}.$$

(Hint: Use the inclusion map from the internal disc to the domain, together with the distance-decreasing property of the Carathéodory metric.)   ■

Exercise. Let

$$U = D \setminus \{0\}$$

be the punctured disc. See Figure 12. This domain does *not* have C^2 boundary. Use the Riemann removable singularities theorem and Cauchy estimates to determine the behavior of the Carathéodory metric near the boundary point 0 of U. Conclude that U is not complete in the Carathéodory metric. What can you say about the Kobayashi metric?   ■

Figure 12.

Corollary 6.2. *Let U be any bounded, finitely connected region in $\mathbb{C}$ (that is, the complement of U has finitely many connected components). Further assume that each boundary component of U is a Jordan curve. Then the Carathéodory metric on U is complete.*

Proof. Since completeness of the Carathéodory metric is a conformal invariant, it is enough to show that U is conformally equivalent to a domain $\widetilde{U}$ having the property that $\partial\widetilde{U}$ has C^2 boundary.

To achieve this end, imagine a planar region with a single hole. Assume that the outer boundary is a Jordan curve, and the boundary of the hole is also a Jordan curve. Fill in the hole to obtain a simply connected region U', and use the Riemann mapping theorem to map U' to the disc. Now unfill the hole. See Figure 13. We now have a region U'' with a circle as the outer boundary and a Jordan curve as the inner boundary.

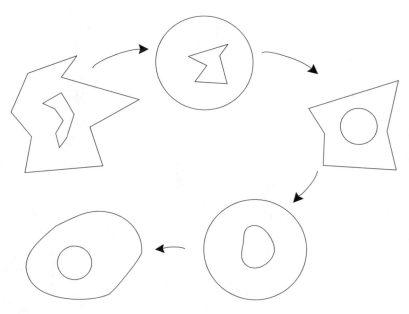

Figure 13.

Let A be the bounded component of the complement of U'' and let p be a point in the interior of A. Do an inversion in p (i.e., $z \mapsto 1/(z - p)$). This maps ∂A to the outer boundary of a new region, and the former (outer) boundary of U'' to the boundary of an interior hole (again see Figure 13). Now fill in the new hole and use the Riemann mapping theorem to map the (filled in) new region to a disc. Finally, perform another inversion (Figure 13). Thus the original hole (that is, A) has been converted to a smooth, circular hole.

If ∂U has k components then k successive applications of inversion and the Riemann mapping theorem, as above, give rise to a conformal mapping of U to a region consisting of the unit disc with $k - 1$ smoothly bounded regions removed. That completes the proof. See [AHL2] for further details.   ∎

We now use the completeness of the Carathéodory metric to prove a version of the Lindelöf principle using the geometric point of view. First we need some terminology.

Let U be a domain with C^2 boundary. Choose a point $P \in \partial U$ and let ν_P be the unit outward normal. If f is a continuous function on U, we say that f has *radial boundary limit* ℓ at P provided that

$$\lim_{r \to 0^+} f(P - r\nu_P) = \ell.$$

If $\alpha > 1$, set

$$\Gamma_\alpha(P) = \{z \in U : |z - P| < \alpha \cdot \text{dist}\,(z, \partial U)\}.$$

Here the expression dist$(z, \partial U)$ denotes Euclidean distance of z to ∂U. We call the domain Γ_α a *Stolz region* or *nontangential approach region* at the point P. See Figure 14. We say that the function f has *nontangential limit* ℓ at P if for each $\alpha > 1$ we have

$$\lim_{\Gamma_\alpha(P) \ni z \to P} f(z) = \ell.$$

Clearly the possession of a nontangential limit at P is a stronger condition than the possession of a radial limit at P. For example, on the unit disc the continuous function

$$f(z) = \frac{y}{1 - x}$$

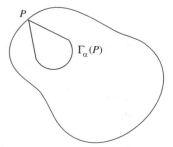

Figure 14.

has radial limit 0 at the boundary point $P = 1 + i0$. However, it does not possess a nontangential limit at P (exercise—try $P_j = (1 - 1/j) + i/j$ for $j = 1, 2, \ldots$). It is therefore surprising that the Lindelöf principle tells us that for bounded holomorphic functions the two notions of boundary limit coincide:

Theorem 7. *Let U_1, U_2 be bounded domains with C^2 boundary and let*

$$f : U_1 \longrightarrow U_2$$

be holomorphic. If $P \in \partial U_1$, $Q \in \partial U_2$, and f has radial limit Q at P, then f has nontangential limit Q at P.

Proof. If z is an element of one of our domains U_j and if $s > 0$ then we let $B(z, s)$ denote the metric ball with center z and radius s in the Carathéodory metric for U_j. For $P \in \partial U_1$, $r_0 > 0$, and $\beta > 0$ fixed, we define

$$\mathcal{M}_\beta(P) = \bigcup_{0 < r < r_0} B(P - r\nu_P, \beta).$$

The estimate

$$F_C^U(z) \approx \frac{C}{\text{dist}(z, \partial U)} \qquad (*)$$

makes it a tedious but not difficult exercise to calculate that the regions $\mathcal{M}_\beta$ are comparable to the regions Γ_α (in this last formula, and in what follows, "dist" means Euclidean distance). In point of fact, suppose that z lies in some $\Gamma_\alpha(p)$, some $p \in \partial U_j$. Let us denote by τ_p the inward normal segment emanating from that boundary point p. Using the estimate (∗), one can then estimate that $\text{dist}_C(z, \tau_p) \leq C \cdot \alpha$. For the converse estimate, assume that $z \notin \Gamma_\alpha(p)$ and the same estimate shows that $\text{dist}_C(z, \tau_p) \geq C \cdot \alpha$.

Thus we see that

$$\lim_{\Gamma_\alpha(P) \ni z \in P} f(z) = \ell, \qquad \forall \alpha > 1$$

iff

$$\lim_{\mathcal{M}_\beta(P) \ni z \to P} f(z) = \ell, \qquad \forall \beta > 0. \tag{∗∗}$$

Thus it is enough to prove (∗∗).

Since

$$\mathcal{M}_\beta(P) = \bigcup_{0 < r < r_0} B(P - r v_P, \beta),$$

the distance-decreasing property of f with respect to the Carathéodory metric implies that

$$f(\mathcal{M}_\beta(P)) \subseteq \bigcup_{0 < r < r_0} B(f(P - r v_P), \beta).$$

Pick $\epsilon > 0$. By the radial limit hypothesis, there is a $\delta > 0$ such that if $0 < t < \delta$, then

$$|f(P - t v_P) - Q| < \epsilon.$$

For such a t, if $z \in B(P - t v_P, \beta)$, then

$$f(z) \in B(f(P - t v_P), \beta).$$

But

$$\text{dist}(f(P - t v_P), \partial U_2) \leq \text{dist}(f(P - t v_P), Q) < \epsilon.$$

Therefore the estimate $(*)$ implies that the metric ball $B(f(P - tv_P), \beta)$ has Euclidean radius not exceeding $C \cdot \epsilon$. Here C depends on β, but β has been fixed once and for all. Thus

$$|f(z) - f(P - tv_P)| < C\epsilon, \qquad \forall z \in B(P - tv_P, \beta).$$

We conclude that

$$|f(z) - Q| \le |f(z) - f(P - tv_P)| + |f(P - tv_P) - Q|$$
$$\le C\epsilon + \epsilon = C'\epsilon.$$

This is the desired conclusion.   ∎

It is not difficult to see that nontangential approach is the broadest possible method for calculating boundary limits of bounded holomorphic functions. See [PRI] for explicit constructions of examples to show that the nontangential approach is the sharp condition. This reference also contains a complete account of the positive results, as well as the history, of the theory of boundary behavior of holomorphic functions.

4. An Application of Completeness: Automorphisms

This section is an introduction to a topic which is usually treated as part of Riemann surface theory. (See [FK] for a nice treatment of the Riemann surface point of view.) We shall not use any of this theory, nor shall we make any reference to Riemann surfaces. Instead, we use geometry.

Definition 1. Let $U \subseteq \mathbb{C}$ be a domain. We let Aut(U) denote the family of conformal self-maps of U (i.e., one-to-one, onto holomorphic functions from U to U).

Proposition 2. *With the binary operation being the composition of functions, the set* Aut(U) *forms a group. We call this group the automorphism group of* U.

Proof. First, the identity mapping id : $U \to U$, id(z) = z, is the group identity. Now if $\phi \in$ Aut(U), then ϕ is one-to-one and onto so that the inverse function ϕ^{-1} makes sense and is holomorphic; it is also one-to-one and onto. And $\phi \circ \phi^{-1} = \phi^{-1} \circ \phi =$ id. Thus the inverse function ϕ^{-1} is the group inverse.

Next, let $\phi, \psi \in$ Aut(U). Then, by definition, $\phi : U \to U$ and $\psi : U \to U$ so certainly $\psi \circ \phi : U \to U$. Since ϕ and ψ are one-to-one, so is the composition $\psi \circ \phi$. Likewise, the composition is onto because each of the individual functions is. Thus Aut(U) is closed under the group operation.

Finally, the group operation is associative because composition of functions is associative. Hence Aut(U) is a group.   ∎

Example 1. Let U be the disc. By the exercise following Theorem 3 of Section 0.2, Aut(U) consists of all maps of the form

$$\zeta \longmapsto \mu \cdot \frac{\zeta - a}{1 - \overline{a}\zeta}$$

where $a, \mu \in \mathbb{C}$, $|a| < 1$, and $|\mu| = 1$.   ∎

Example 2. Let $0 < r < R < \infty$ and let

$$A_{r,R} = \{z : r < |z| < R\}.$$

Then Aut($A_{r,R}$) consists only of the rotations

$$z \longmapsto \mu \cdot z \quad \text{(for } \mu \text{ a unimodular constant)},$$

the reflection

$$\sigma : z \longmapsto \frac{R \cdot r}{z},$$

and compositions of these two types of functions (see Theorem 1 of Section 4.6).

Clearly each of these two types of functions is an element of $\mathrm{Aut}(A_{r,R})$. For the converse, let $\phi \in \mathrm{Aut}(A_{r,R})$. Since ϕ^{-1} is continuous, ϕ is proper. This means that the inverse image of any compact set under ϕ is compact. Taking contrapositives, we see that if $z_j \in A_{r,R}$ satisfies $z_j \to \partial A_{r,R}$, then $\phi(z_j) \to \partial A_{r,R}$. Thus we have confirmed that the boundary goes to the boundary (and either Carathéodory's theorem, or the maximum principle, or Brouwer invariance of domain are other standard ways to see this assertion). A tricky but elementary topological argument (see [GRK, pp. 243–244]) then shows that either the inner boundary of $A_{r,R}$ is mapped to the inner boundary of $A_{r,R}$ or else the inner boundary of $A_{r,R}$ is mapped to the outer boundary. After applying an inversion, we may suppose the former.

Notice that the Schwarz reflection principle implies that ϕ continues analytically past $\partial A_{r,R}$. We claim that ϕ must be a rotation. We may reflect ϕ across both boundaries of $A_{r,R}$ to obtain an automorphism of the larger annulus $\{z : r^2/R < |z| < R^2/r\}$. Continuing this reflection process countably many times, we can analytically continue ϕ to an automorphism of $\mathbb{C}' \equiv \mathbb{C} \setminus \{0\}$. By the Riemann removable singularities theorem, ϕ continues analytically to an automorphism of $\mathbb{C}$ which fixes the origin. Thus there is a constant α such that $\phi(z) \equiv \alpha z$. Since ϕ is the continuation of an automorphism of a proper annulus $A_{r,R}$, it must be that α has modulus 1. Therefore ϕ is a rotation as claimed. ∎

Given a domain U, we equip $\mathrm{Aut}(U)$ with the topology of uniform convergence on compact sets. That is,

$$\mathrm{Aut}(U) \ni \phi_j \longrightarrow \phi \in \mathrm{Aut}(U)$$

provided that $\phi_j \to \phi$ normally. We are particularly interested in the question of when the topological group $\mathrm{Aut}(U)$ is *compact*. Given that we have defined the topology sequentially, it is natural to consider sequential compactness: $\mathrm{Aut}(U)$ is compact if and only if, whenever $\{\phi_j\} \subseteq \mathrm{Aut}(U)$, then there is a subsequence $\{\phi_{j_k}\}$ that is convergent (uniformly on compact subsets of U) *to an element of* $\mathrm{Aut}(U)$.

Example 3. Let $0 < r < R < \infty$ and let

$$A_{r,R} = \{z : r < |z| < R\}$$

be the corresponding annulus. Then $\text{Aut}(A_{r,R})$ is compact. To see this, let α_j be a sequence of automorphisms of $A_{r,R}$. Now each of the α_j is either a rotation, the reflection $\sigma(z) = r \cdot R/z$, or a composition of the two. Thus infinitely many of the α_j are of the same one of these three types. If infinitely many are equal to the reflection σ, then a subsequence consisting of these elements (this will be a constant sequence) converges to σ. If infinitely many are rotations, then let us write these as

$$\alpha_{j_k}(z) = \mu_{j_k} \cdot z,$$

where the μ_{j_k} are unimodular constants. Since the unit circle is compact, there is a subsequence—call it μ_m—which converges to a unimodular constant μ_0. But then the corresponding automorphisms

$$\alpha_m(z) = \mu_m z$$

converge normally to the limit automorphism $\alpha_0(z) = \mu_0 z$. The third possibility is that infinitely many of the α_j consist of a rotation followed by the reflection σ. Write these as

$$\alpha_{j_k}(z) = \sigma(\mu_{j_k} z),$$

with the μ_{j_k} being unimodular constants. We extract a convergent subsequence $\{\mu_m\}$ with limit μ_0 and define

$$\alpha_0(z) = \sigma(\mu_0 z).$$

Then the subsequence

$$\alpha_m(z) = \sigma(\mu_m z)$$

converges normally to $\alpha_0(z)$.

The three possibilities having been covered, we conclude that $\text{Aut}(A_{r,R})$ is compact. ∎

Remark. It is an interesting fact that if the connectivity of a planar domain is finite but greater than two then the automorphism group of the domain is a finite group (see [FK] or [HEI]). That is to say, a domain with at least two holes, but finitely many, has automorphism group that is a finite group. It is an open problem to determine *which* finite groups arise as automorphism groups of planar domains. ∎

Example 4. Let $U = D$, the unit disc. Then the functions

$$\beta_j(\zeta) = \frac{\zeta + (1 - 1/j)}{1 + (1 - 1/j)\zeta}$$

are elements of Aut(D). However, $\{\beta_j\}$ converges normally to the function

$$\beta_0(\zeta) \equiv 1.$$

Notice that $\beta_0 \notin$ Aut(D)—in fact it is a constant function. Therefore Aut(D) is not compact. ∎

Remark. Notice that, at least when U is bounded, the issue of whether Aut(U) is compact is not a question of whether $\{\phi_j\} \subseteq$ Aut(U) has a subsequence that converges: by normal families this is essentially automatic. Rather, the issue is *whether the subsequence will converge to an element of Aut(U)*. It is this last phenomenon which fails for the disc, but works for the annulus. ∎

The principal result of this section is that the disc is characterized by the non-compactness of its automorphism group. More precisely, we have

Theorem 3. *Let $U \subseteq \mathbb{C}$ be a bounded domain with C^2 boundary. If Aut(U) is non-compact then U is conformally equivalent to the unit disc.*

We prove this theorem with a sequence of lemmas, each of which has intrinsic interest. Take a moment to review the notions of C^2 boundary and C^k boundary which were discussed in Definitions 2 and 3 of Section 3.

Lemma 4. *Let $U \subseteq \mathbb{C}$ be bounded. The group Aut(U) is compact if and only if, for each $P \in U$, there is a compact $K_P \subseteq U$ such that $\phi(P) \in K_P$ for all $\phi \in$ Aut(U).*

Proof. Assume that Aut(U) is compact. Fix $P \in U$. If there is no set K_P as claimed then there exist $\phi_j \in$ Aut(U) such that $\phi_j(P) \to w \in \partial U$, some w. But U is bounded so that $\{\phi_j\}$ is a normal family; thus there is a subsequence ϕ_{j_k} and a holomorphic limit function ϕ_0 such that

$$\phi_{j_k} \longrightarrow \phi_0$$

normally.

Notice that the image of each ϕ_j lies in U, hence the image of ϕ_0 lies in the *closure* $\overline{U}$ of U. If ϕ_0 is nonconstant then it satisfies the open mapping principle. But

$$\phi_0(P) = \lim_{k \to \infty} \phi_{j_k}(P) = w,$$

hence the image of ϕ_0 contains the accumulation point $w \in \partial U$, so it contains a neighborhood of w. This is impossible because w is in the boundary of the image of ϕ_0. Therefore ϕ_0 must be constantly equal to w; thus $\phi_0 \notin$ Aut(U). The sequence ϕ_{j_k} therefore violates the compactness of Aut(U). We conclude that K_P must exist.

For the converse, fix $P \in U$ and let K_P be the corresponding compact set in U whose existence we assume. Let $\{\phi_j\} \subseteq$ Aut(U) be any sequence. Since U is bounded, there is a normally converging subsequence ϕ_{j_k} with holomorphic limit function ϕ_0. As in the first half of the proof, if the image of ϕ_0 contains any boundary point w then ϕ_0 must be constantly equal to w. But the image of P under ϕ_0 must lie in K_P, so this possibility is ruled out. We conclude that the image of ϕ_0 lies in U.

Next notice that each ϕ_{j_k} has an inverse ψ_{j_k}. Passing to another subsequence, we may suppose that the ψ_{j_k} converge to a limit function ψ_0. For convenience, we denote this final subsequence by ψ_m, corresponding to the automorphisms ϕ_m. Just as for ϕ_0, we can be sure that the image of ψ_0 lies in U. As in the proof of Theorem 5 of Section 2, ψ_0 is non-constant.

Now we have

$$z \equiv \lim_{m \to \infty} \phi_m \circ \psi_m(z) = \phi_0 \circ \psi_0(z).$$

Since $\mathbf{i}(z) \equiv z$ is onto, so is ϕ_0. Also, as in the proof of Theorem 5 in Section 2, the image of ψ_0 is open, closed, and nonempty. Therefore ψ_0 is surjective. Since $\mathbf{i}(z)$ is injective, it now follows that ψ_0 is injective. Therefore $\psi_0 \in \text{Aut}(U)$ and it follows that $\phi_0 \in \text{Aut}(U)$. We conclude that

$$\text{Aut}(U) \ni \phi_m \longrightarrow \phi_0 \in \text{Aut}(U),$$

and $\text{Aut}(U)$ is compact. ∎

Remark. It may be noted that, in the proof of the converse direction of Lemma 4, only one compact set K_P was needed. ∎

Lemma 5. *Let $U \subseteq \mathbb{C}$ be a bounded domain with C^2 boundary. Suppose that $P \in U$, $\{\phi_j\}$ are holomorphic maps from U to U, and*

$$\phi_j(P) \longrightarrow w \in \partial U.$$

If K is compact in U and V is a neighborhood of w, then there exists a positive number J such that if $j \geq J$ then

$$\phi_j(K) \subseteq V.$$

See Figure 1.

Proof. Since U, when equipped with the Carathéodory metric, is a metric space and since K is compact, there is a positive number R such that

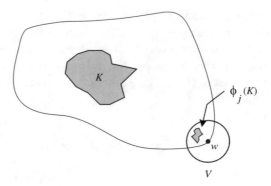

Figure 1.

the metric ball $B(P, R)$ contains K (see Proposition 1 in Section 3). Let $Q_j = \phi_j(P)$. Since each ϕ_j is distance-decreasing in the Carathéodory metric, it follows that $\phi_j(B(P, R)) \subseteq B(Q_j, R)$. We claim that there is a positive J such that whenever $j \geq J$ then $B(Q_j, R) \subseteq V$. Assuming the claim, we would then have

$$\phi_j(K) \subseteq \phi_j(B(P, R)) \subseteq B(Q_j, R) \subseteq V,$$

as required.

To prove the claim, recall from the proof of Theorem 6 in the last section that (because the Carathéodory metric on U is complete) the Euclidean radii of the metric balls $B(Q_j, R)$ must tend to 0. Choose $\epsilon > 0$ such that the Euclidean disc of center w and radius 2ϵ lies in V. We select J so large that when $j > J$, then both the Euclidean distance of Q_j to w is less than ϵ and the Euclidean radius of $B(Q_j, R)$ is less than ϵ. The claim now follows from the triangle inequality. ∎

Proof of Theorem 3. If Aut(U) is not compact then, by Lemma 4, there is a sequence $\phi_j \in$ Aut(U) and a $P \in U$ such that

$$\phi_j(P) \longrightarrow w \in \partial U,$$

for some $w \in \partial U$.

Let

$$\gamma : [0, 1] \longrightarrow U$$

be any continuous, closed curve in U. Since ∂U is C^2 there is a neighborhood V of w such that $U \cap V$ is simply connected (see Figure 2—the existence of the interior osculating circle, or the tubular neighborhood, provided by Proposition 3 of Section 3 makes this assertion clear).

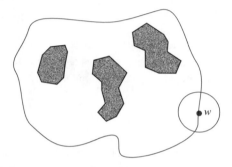

Figure 2.

Let

$$K = \{\gamma(t) : 0 \le t \le 1\}.$$

Then K is compact. By Lemma 5, there is a $J \ge 0$ such that $j \ge J$ implies $\phi_j(K) \subseteq U \cap V$. Thus $\phi_j \circ \gamma$ is a continuous, closed curve in $U \cap V$. The simple connectivity of $U \cap V$ implies that $\phi_j \circ \gamma$ may be continuously deformed to the point $\phi_j \circ \gamma(0)$; that is, there is a homotopy

$$\Psi : [0, 1] \times [0, 1] \longrightarrow U \cap V$$

such that

$$\Psi(0, t) = \phi_j \circ \gamma(t), \qquad \forall t \in [0, 1]$$

and

$$\Psi(1, t) = \phi_j \circ \gamma(0), \qquad \forall t \in [0, 1].$$

But then

$$(\phi_j)^{-1} \circ \Psi$$

is a homotopy of the curve γ to the point $\gamma(0)$. It follows that U is simply connected. By the Riemann mapping theorem, U is conformally equivalent to the disc. ∎

Remark. It is possible to prove this theorem without resort to the Riemann mapping theorem. Instead one uses a delicate approximation argument on a decreasing sequence of boundary neighborhoods $V_j \cap U$ of w to prove that

$$F_C^U(P) = F_K^U(P).$$

It then follows from Theorem 5 of Section 2 that U is conformally equivalent to the disc. In the same vein, one can actually prove the Riemann mapping theorem using F_C^U and F_K^U. Since the ideas involved are rather complicated and would take us far afield, we omit them. ∎

The notion of compact/non-compact automorphism group is a bit abstract, so we conclude this section with an interpretation of Theorem 3 which is more concrete. Let us say that Aut(U) *acts transitively* on the domain U if for any P, Q in U there is an element $\phi \in$ Aut(U) such that $\phi(P) = Q$. In other words, the automorphism group acts transitively if it is sufficiently rich with elements that it can move any point of U to any other.

Example 5. Let $U = D$, the unit disc. If P is any element of D then the automorphism

$$\phi_P(\zeta) \equiv \frac{\zeta - P}{1 - \overline{P}\zeta}$$

has the property that $\phi_P(P) = 0$. Now if P and Q are arbitrary elements of D then $(\phi_Q)^{-1} \circ \phi_P$ is an automorphism of D that maps P to Q. Thus the automorphism group of the disc *does* act transitively.

■

Example 6. Let $A_{r,R}$ be the annulus

$$\{z : r < |z| < R\}.$$

Then $\mathrm{Aut}(A_{r,R})$ does *not* act transitively on $A_{r,R}$. In fact if $P, Q \in A_{r,R}$, if $|P| \neq |Q|$, and if $|P| \neq r \cdot R/|Q|$ then Example 2 shows that there is no automorphism of $A_{r,R}$ which takes P to Q. ■

We can now give a consequence of Theorem 3, using the language of transitivity.

Corollary 5.1. *Let U be a bounded domain with C^2 boundary. The group $\mathrm{Aut}(U)$ acts transitively on U if and only if U is conformally equivalent with the disc.*

Proof. Fix a point P in U. If $\mathrm{Aut}(U)$ acts transitively on U, then the point P can be moved to any other point of U by some element of $\mathrm{Aut}(U)$. Thus $\{\phi(P) : \phi \in \mathrm{Aut}(U)\}$ can be contained in no compact set $K_P \subseteq U$. By Lemma 4, $\mathrm{Aut}(U)$ is non-compact. By the theorem, U must be conformally equivalent to the disc.

The converse is obvious from our explicit knowledge of the automorphisms of the disc. ■

It is natural to wonder whether the hypothesis of the C^2 boundary is really necessary for the results presented in this section. The answer is that *some* regularity of the boundary is required, as the next example illustrates.

Example 7. Define the mapping

$$\phi(z) = \frac{z + 1/2}{1 + (1/2)z}.$$

We know that ϕ is an automorphism of the disc. Define $\phi^2 = \phi \circ \phi$, $\phi^3 = \phi \circ \phi \circ \phi$, etc. With ϕ^{-1} denoting the inverse of ϕ, we set $\phi^{-2} = \phi^{-1} \circ \phi^{-1}$, $\phi^{-3} = \phi^{-1} \circ \phi^{-1} \circ \phi^{-1}$, etc. We let $\phi^0(z) \equiv z$.

For $j \in \mathbb{Z}$ we define

$$\overline{d}_j = \{\phi^j(z) : |z| \leq 1/10\}.$$

Finally, let

$$U = D(0, 1) \setminus \bigcup_{j \in \mathbb{Z}} \overline{d}_j.$$

The domain U is displayed in Figure 3.

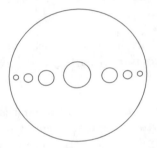

Figure 3.

It is not difficult to verify that $\mathrm{Aut}(U)$ consists precisely of the maps ϕ^j, $j \in \mathbb{Z}$, together with the map $z \mapsto -z$. Thus $\mathrm{Aut}(U)$ is non-compact (exercise). Yet U is certainly not conformally equivalent to the disc.

Notice that the boundary of U is not C^2, in the sense that it is impossible to write $U = \{z \in \mathbb{C} : \rho(z) < 0\}$ for any C^2 defining function ρ with $\nabla \rho \neq 0$ on ∂U; so there is no contradiction with Theorem 3. ∎

Exercise. Determine the automorphism group of the punctured disc. Is it compact or not? ∎

5.   Hyperbolicity and Curvature

In this section we relate the concepts of curvature and normal families to the property of nondegeneracy of the Kobayashi metric. A consequence is that we can prove that the plane and $\mathbb{C} \setminus \{0\}$ cannot support a metric of strictly negative curvature. Thus we see, from a geometric point of view, why Picard's theorem must specify the omission of *two* points of the range of an entire function in order for that function to be necessarily constant.

We begin with an elegant generalization of the fact that the Carathéodory metric is never greater than the Kobayashi metric. Recall here that ρ denotes the Poincaré metric on the disc D.

Proposition 1.   *If $U \subseteq \mathbb{C}$ is a domain which is equipped with a metric σ and if every holomorphic map $f : D \to U$ is distance-decreasing from (D, ρ) to (U, σ) then*

$$\sigma \leq F_K^U .$$

Proof.   Abbreviate F_K^U by the symbol μ. Let $z \in U$. If $\phi : D \to U$ and $\phi(0) = z$ then, by hypothesis,

$$\phi^* \sigma(0) \leq \rho(0)$$

or

$$\frac{1}{|\phi'(0)|} \geq \frac{\sigma(z)}{\rho(0)} = \sigma(z).$$

Taking the infimum over all ϕ we conclude, for any $\xi \in \mathbb{C}$, that

$$\|\xi\|_{\mu,z} = \inf_{\phi \in (U,D)_z} \frac{|\xi|}{|\phi'(0)|} \geq |\xi| \cdot \sigma(z) = \|\xi\|_{\sigma,z},$$

as desired.   ∎

Definition 2.   A domain $U \subseteq \mathbb{C}$ is called *hyperbolic* if the Kobayashi distance on U is actually a distance, that is, if

$$d_{\text{Kob}}(P, Q) > 0$$

whenever P and Q are distinct points of U.

Proposition 3. *Let $U \subseteq \mathbb{C}$ be a domain equipped with a C^2 metric σ which has the property that its curvature $\kappa = \kappa_\sigma$ satisfies*

$$\kappa \le -B < 0$$

for some positive constant B. Then U is hyperbolic.

Proof. Let f be any map of the disc D into U. By Ahlfors's version of Schwarz's lemma, we have that

$$f^* \sigma \le \frac{\sqrt{4}}{\sqrt{B}} \rho,$$

where ρ is the Poincaré metric on the disc. Replace σ by $\tilde{\sigma} \equiv (\sqrt{B}/\sqrt{4}) \cdot \sigma$. Then our inequality is

$$f^* \tilde{\sigma} \le \rho.$$

Thus f is distance-decreasing from (D, ρ) to $(U, \tilde{\sigma})$. By Proposition 1, we may conclude that

$$\tilde{\sigma} \le F_K^U$$

or

$$C \cdot \sigma \le F_K^U,$$

where $C = \sqrt{B}/\sqrt{4}$ is a positive constant. Thus the Kobayashi distance is bounded from below by a positive constant times the σ-distance. Since σ is assumed to be nondegenerate, so is F_K^U. ∎

Corollary 3.1. *Neither the plane $\mathbb{C}$ nor the punctured plane $\mathbb{C} \setminus \{0\}$ possesses a metric of strictly negative curvature.*

Proof. Neither $\mathbb{C}$ nor $\mathbb{C} \setminus \{0\}$ is hyperbolic. ∎

Corollary 3.2. *If $U = \mathbb{C} \setminus \{P_1, \dots, P_k\}$, with the P_j's distinct and $k \ge 2$, then U is hyperbolic.*

Proof. As we saw in Chapter 2, such a U has a metric of strictly negative curvature.

∎

Remark. An alternative proof of Corollary 3.2 may be obtained by noticing that a domain U which satisfies the hypothesis will imbed into $\mathbb{C} \setminus \{0, 1\}$ and then using the distance-decreasing property of the Kobayashi metric.

∎

Now we will connect hyperbolicity with the concept of normal families. We first need some new terminology.

Definition 4. A domain U is *taut* if the family $\mathcal{F}$ of holomorphic functions from the disc D into U is normal.

Example 1. The disc D is taut. For if $\mathcal{F}$ is the family of holomorphic functions from D to D then the elements of $\mathcal{F}$ are uniformly bounded by 1. So any sequence of elements $\{f_j\} \subseteq \mathcal{F}$ has a subsequence $\{f_{j_k}\}$ which converges to a limit function f_0. The function f_0 is certainly holomorphic, and there are two possibilities: i) the image of f_0 contains a boundary element w, in which case the open mapping principle implies that f_0 is constant (this is the compactly divergent case); (ii) the image of f_0 lies entirely in the interior of D, in which case $f_0 \in \mathcal{F}$ (this is the convergent case).

∎

Example 2. The domain $U = \mathbb{C} \setminus \{0\}$ is not taut. The functions

$$f_n(z) = e^{nz}$$

map the disc into U but there is neither a convergent subsequence of $\{f_n\}$ nor a compactly divergent one (the origin is mapped to 1 by every f_n but $f_n(1/2) \to \infty$ as $n \to \infty$).

∎

Proposition 5. *A planar domain U is hyperbolic if and only if it is taut.*

Proof. We know that if U is hyperbolic then $\mathbb{C} \setminus U$ must contain at least two points. By the general version of Montel's theorem in Section 2.4, U must therefore be taut.

   Conversely, assume that U is taut. Then, by Example 2, $\mathbb{C} \setminus U$ must contain at least two points. But then U is hyperbolic.   ∎

Remark. One theme which comes through in this section is that the only enemies are $\mathbb{C}$ and $\mathbb{C} \setminus \{0\}$. As soon as a domain excludes at least two points, then it has all relevant properties. In the theory of Riemann surfaces it also holds that a domain is taut if and only if it is hyperbolic (see [FK]). However, in several complex variables, matters are no longer so simple. A detailed discussion of the relationship between tautness and hyperbolicity appears in [KOB2, p. 240]. See also [KOB1].

Exercise. The domain

$$U = \{z \in \mathbb{C} : |z| < 1\} \setminus \{x + i0 : 0 \le x < 1\}$$

is taut and hyperbolic.   ∎

CHAPTER **4**

Introduction to the Bergman Theory

0. Introductory Remarks

It is a remarkable fact—discovered by Stefan Bergman in 1927—that a bounded domain Ω in $\mathbb{C}$ or in $\mathbb{C}^n$ can be equipped with a "canonical" reproducing kernel. [Here we use the phrase "reproducing kernel" to mean a function $k(z, \zeta)$ of two variables—like the familiar Cauchy kernel—with the property that integration against a holomorphic function f produces the value of f at z. In one complex variable such a formula could take the form

$$f(z) = \int_{\partial\Omega} k(z, \zeta) f(\zeta) \, d\zeta$$

or

$$f(z) = \iint_{\Omega} k(z, \zeta) f(\zeta) \, d\xi \, d\eta$$

when $\zeta = \xi + i\eta$. Such a formula means, in effect, that the kernel k contains valuable information about holomorphic functions on Ω.] In one complex variable, the existence of a reproducing kernel is not great news. The Cauchy kernel, after all, works for any domain. But in Bergman's time there were no reproducing kernels in several complex variable (except on very special domains like the ball and the polydisc—see Chapter 5 below), and his goal was to produce a construction of a kernel that worked on virtually any domain.

137

A particularly remarkable byproduct of Bergman's construction is that his kernel can in turn be used to produce a "canonical invariant metric." This metric has many special properties: it is smooth (indeed real analytic), it is Hermitian, it is Kähler, and it is invariant under biholomorphic mappings. As a result, this "Bergman metric" has been of great interest to geometers for more than three quarters of a century.

In the present chapter we shall give a very brief, and occasionally sketchy, introduction to the Bergman kernel and metric. The basic elements of the Bergman theory rely on Hilbert space ideas, and we do not wish to burden the reader with this extra formalism. So we shall steer clear of most of Hilbert space, at the expense of not being able to provide all proofs in detail. Still, we believe that this chapter will offer many rewards to the interested reader. We shall be able to calculate the Bergman kernel and metric on at least some domains, and we shall be able to verify their invariance explicitly. We shall provide means (in fact three different methods) for constructing the Bergman kernel. And we will apply the new Bergman metric to a problem in conformal mapping.

We encourage the reader to study [RUD3] for details on Hilbert space theory. The books [GRK], [KR1], [FUK], [EPS] provide further details of the Bergman theory.

1. Bergman Basics

The Bergman theory centers on a special function space. Let $\Omega \subseteq \mathbb{C}$ be a bounded domain. Now define the *Bergman space*

$$A^2(\Omega) = \{f \text{ holomorphic on } \Omega : \iint_\Omega |f(z)|^2 \, dA(z) < \infty\}.$$

Here dA is the usual area measure on Ω. In other words, $dA = dx \, dy$. We define a *norm* on $A^2(\Omega)$ by

$$\|f\|_{A^2(\Omega)} \equiv \left[\iint_\Omega |f(z)|^2 \, dA(z)\right]^{1/2}.$$

This norm satisfies the standard properties which we would expect of such an operation:

1. $\|f\|_{A^2(\Omega)} \geq 0$;
2. $\|f + g\|_{A^2(\Omega)} \leq \|f\|_{A^2(\Omega)} + \|g\|_{A^2(\Omega)}$;
3. $\|cf\|_{A^2(\Omega)} = |c| \cdot \|f\|_{A^2(\Omega)}$ for any complex constant c.

There is an inner product on $A^2(\Omega)$ which is naturally associated to this norm. Namely, for $f, g \in A^2(\Omega)$,

$$\langle f, g \rangle_{A^2(\Omega)} = \langle f, g \rangle = \iint_\Omega f(z)\overline{g(z)} \, dA(z).$$

Naturally we could give similar definitions on a bounded domain $\Omega \subseteq \mathbb{C}^n$, with area measure dA replaced by volume measure dV. The theory would go through without any change. For simplicity, however, we shall concentrate in this chapter on domains in the complex plane.

It is important to observe that, as long as Ω is a bounded domain, the space $A^2(\Omega)$ is certainly non-empty. For example, any holomorphic polynomial (restricted to Ω) will be in $A^2(\Omega)$. If Ω is simply connected and $P \in \partial\Omega$ then $g(z) = 1/\sqrt{z - P}$ (using the principal branch of square root, as we may) will be in $A^2(\Omega)$.

Now the *Bergman kernel* $K(z, \zeta)$ is a function on the set $\Omega \times \Omega$ with these properties:

1. For each fixed $\zeta \in \Omega$, $K(\cdot, \zeta) \in A^2(\Omega)$.
2. For each fixed $z \in \Omega$, $\overline{K(z, \cdot)} \in A^2(\Omega)$.
3. For $z, \zeta \in \Omega$, $K(z, \zeta) = \overline{K(\zeta, z)}$.
4. If $f \in A^2(\Omega)$ and $z \in \Omega$ then

$$f(z) = \iint_\Omega K(z, \zeta) f(\zeta) \, dA(\zeta).$$

Observe that properties (1) and (3) already imply (2). Also (2) and (3) imply (1). But it is useful to have these attributes all laid out explicitly. It is important to note that properties (1)–(4) characterize the

Bergman kernel: Any function $k(z, \zeta)$ satisfying (1)–(4) will satisfy $k = K$. This characterization will prove useful below.

One can use abstract Hilbert space methods to prove the existence of the Bergman kernel. We instead will present, in Section 3, three different methods for constructing the Bergman kernel. The third method in effect proves the existence of the Bergman kernel, for it expresses the Bergman kernel in terms of the well-known Green's function.

Property (4) is the reproducing property of the Bergman kernel. Compare, for example, the Cauchy integral formula for holomorphic functions. At first blush, we might suspect that the Bergman kernel *is* the Cauchy kernel. But bear in mind that the Bergman kernel is a kernel for integration with respect to area while the Cauchy kernel is a kernel for complex line integration. It turns out that this information alone forces the two kernels to be different.

Another important point to notice is that the Cauchy kernel $C(z, \zeta) = 1/(\zeta - z)$ is the *same* for all domains. In fact, Bergman kernels for different domains are (generically) different.

2. Invariance Properties of the Bergman Kernel

Suppose that Ω_1, Ω_2 are bounded domains in $\mathbb{C}$. Let $\phi : \Omega_1 \to \Omega_2$ be a conformal mapping. We might hope that the Bergman kernel K_{Ω_1} for Ω_1 and the Bergman kernel K_{Ω_2} for Ω_2 are related by way of ϕ.

To establish such a fact, we first need a technical result about the way that area transforms under conformal mapping.

Lemma 1. (Lusin). *Let* $\phi : \Omega_1 \to \Omega_2$ *be a conformal mapping of domains. Then*

$$\text{area}\,(\Omega_2) = \iint_{\Omega_2} 1 \, dA(z) = \iint_{\Omega_1} |\phi'(\zeta)|^2 \, dA(\zeta).$$

Proof. The proof amounts to writing the standard change-of-variables formula in complex notation. Write ϕ in terms of its real and imaginary parts as

$$\phi(\zeta) = \phi_1(\zeta) + i\phi_2(\zeta).$$

Also write $\zeta = \tau + i\eta$. Then the usual Jacobian matrix determinant Jac ϕ of ϕ is

$$\det \begin{pmatrix} \dfrac{\partial \phi_1}{\partial \tau} & \dfrac{\partial \phi_1}{\partial \eta} \\[3mm] \dfrac{\partial \phi_2}{\partial \tau} & \dfrac{\partial \phi_2}{\partial \eta} \end{pmatrix} = \frac{\partial \phi_1}{\partial \tau} \cdot \frac{\partial \phi_2}{\partial \eta} - \frac{\partial \phi_2}{\partial \tau} \cdot \frac{\partial \phi_1}{\partial \eta}.$$

We may apply the Cauchy-Riemann equations twice to convert this last expression to

$$\frac{\partial \phi_1}{\partial \tau} \cdot \frac{\partial \phi_1}{\partial \tau} + \frac{\partial \phi_2}{\partial \tau} \cdot \frac{\partial \phi_2}{\partial \tau} = \left| \frac{\partial \phi}{\partial \tau} \right|^2 = \left| \frac{\partial \phi}{\partial \zeta} \right|^2 = |\phi'(\zeta)|^2.$$

In the last step we have used the fact that $\partial \phi / \partial \tau = \partial \phi / \partial \zeta = \phi'$ for a holomorphic function ϕ.

   In conclusion,

$$\text{area}\,(\Omega_2) = \iint_{\Omega_2} 1\, dA(z) = \iint_{\Omega_1} \det \text{Jac}\, \phi(\zeta)\, dA(\zeta)$$

$$= \iint_{\Omega_1} |\phi'(\zeta)|^2\, dA(\zeta),$$

as the lemma asserts. ∎

   Now we have

Theorem 2. *If Ω_1, Ω_2, ϕ are as in the introductory paragraph, then, for $z, \zeta \in \Omega_1$,*

$$K_{\Omega_1}(z, \zeta) = \phi'(z) \cdot K_{\Omega_2}(\phi(z), \phi(\zeta)) \cdot \overline{\phi'(\zeta)}.$$

Proof. Let $f \in A^2(\Omega_1)$. Then (applying the change of variable $\xi = \phi(\zeta)$—valid because ϕ and ϕ^{-1} are univalent and continuously differentiable)

$$\iint_{\Omega_1} \phi'(z) \ \cdot \ K_{\Omega_2}(\phi(z), \phi(\zeta)) \cdot \overline{\phi'(\zeta)} f(\zeta) \, dA(\zeta)$$

$$= \iint_{\Omega_2} \phi'(z) \cdot K_{\Omega_2}(\phi(z), \phi(\phi^{-1}(\xi)))$$

$$\times \overline{\phi'(\phi^{-1}(\xi))} f(\phi^{-1}(\xi)) \left| [\phi^{-1}]'(\xi) \right|^2 dA(\xi)$$

$$= \iint_{\Omega_2} \phi'(z) \cdot K_{\Omega_2}(\phi(z), \xi)$$

$$\times \overline{\phi'(\phi^{-1}(\xi))} f(\phi^{-1}(\xi))$$

$$\times \frac{1}{\phi'(\phi^{-1}(\xi)) \overline{\phi'(\phi^{-1}(\xi))}} dA(\xi)$$

$$= \iint_{\Omega_2} \phi'(z) \cdot K_{\Omega_2}(\phi(z), \xi)$$

$$\times \left[f(\phi^{-1}(\xi)) \cdot \frac{1}{\phi'(\phi^{-1}(\xi))} \right] dA(\xi)$$

$$\equiv \phi'(z) \cdot \iint_{\Omega_2} K_{\Omega_2}(\phi(z), \xi) \cdot g(\xi) \, dA(\xi),$$

where $g(\xi) = f(\phi^{-1}(\xi)) \cdot \frac{1}{\phi'(\phi^{-1}(\xi))}$.

It is easy to verify, using a change of variable, that $g \in A^2(\Omega_2)$. As a result, the Bergman kernel K_{Ω_2} for Ω_2 will reproduce g. We conclude that

$$\iint_{\Omega_1} \phi'(z) \ \cdot \ K_{\Omega_2}(\phi(z), \phi(\zeta)) \cdot \overline{\phi'(\zeta)} f(\zeta) \, dA(\zeta)$$

$$= \phi'(z) \cdot g(\phi(z)) = f(z).$$

We have verified that the kernel

$$k(z, \zeta) \equiv \phi'(z) \cdot K_{\Omega_2}(\phi(z), \phi(\zeta)) \cdot \overline{\phi'(\zeta)}$$

satisfies the reproducing property (4) for the Bergman kernel.

　　　The kernel $\phi'(z) \cdot K_{\Omega_2}(\phi(z), \phi(\zeta)) \cdot \overline{\phi'(\zeta)}$ plainly satisfies the conjugate symmetric property (3) of the Bergman kernel for Ω_1. We also note that k is holomorphic in the z-variable and conjugate holomorphic in the ζ-variable. Properties (1) and (2) may be checked by a change of variables just like the one we performed to prove the reproducing property. We omit the details.

　　　We may conclude, by the uniqueness of the Bergman kernel, that

$$\phi'(z) \cdot K_{\Omega_2}(\phi(z), \phi(\zeta)) \cdot \overline{\phi'(\zeta)} = K_{\Omega_1}(z, \zeta).$$

This is what we wished to prove.　　　　　　　　　　　　　　　■

3.　Calculation of the Bergman Kernel

The usual construction of the Bergman kernel is rather abstract, relying as it does on the Riesz representation theorem of Hilbert space theory (see [RU3] for the details). But this gives us little leverage for actually calculating the kernel. In the present section we shall present three different techniques for constructing the Bergman kernel of the unit disc. Theorem 2 of Section 2 then gives, at least in principle, the means to calculate the Bergman kernel for other simply connected domains.

3.1　Construction of the Bergman Kernel for the Disc by Conformal Invariance

Let $D \subseteq \mathbb{C}$ be the unit disc. First we notice that, if either $f \in A^2(D)$ or $\overline{f} \in A^2(D)$, then

$$f(0) = \frac{1}{\pi} \iint_D f(\zeta) \, dA(\zeta). \qquad (*)$$

　This is the standard area form of the mean value property for holomorphic or harmonic functions.

Of course the constant function $u(z) \equiv 1$ is in $A^2(D)$, so it is reproduced by integration against the Bergman kernel. Hence, for any $w \in D$,

$$1 = u(w) = \iint_D K(w, \zeta) u(\zeta) \, dA(\zeta) = \iint_D K(w, \zeta) \, dA(\zeta)$$

or

$$\frac{1}{\pi} = \frac{1}{\pi} \iint_D K(w, \zeta) \, dA(\zeta).$$

By $(*)$, we may conclude that

$$\frac{1}{\pi} = K(w, 0)$$

for any $w \in D$.

Now, for $a \in D$ fixed, consider the Möbius transformation

$$h(z) = \frac{z - a}{1 - \overline{a}z}$$

that we studied in Section 1.4. We know that

$$h'(z) = \frac{1 - |a|^2}{(1 - \overline{a}z)^2}.$$

We may thus apply Theorem 2 of Section 4.2 with $\phi = h$ to find that

$$\begin{aligned}
K(w, a) &= h'(w) \cdot K(h(w), h(a)) \cdot \overline{h'(a)} \\
&= \frac{1 - |a|^2}{(1 - \overline{a}w)^2} \cdot K(h(w), 0) \cdot \frac{1}{1 - |a|^2} \\
&= \frac{1}{(1 - \overline{a}w)^2} \cdot \frac{1}{\pi} \\
&= \frac{1}{\pi} \cdot \frac{1}{(1 - w\overline{a})^2}.
\end{aligned}$$

Thus, assuming the existence of the Bergman kernel of the disc, we have derived a closed-form formula for it.

3.2 Construction of the Bergman Kernel by Means of an Orthonormal System

One of the most standard means of calculating the Bergman kernel is the following lemma:

Lemma 1. *Let $\{\phi_j\}$ be a complete orthonormal system for $A^2(\Omega)$. Then the function*

$$k(z, \zeta) \equiv \sum_{j=1}^{\infty} \phi_j(z) \cdot \overline{\phi_j(\zeta)} \qquad (\star)$$

is in fact the Bergman kernel for Ω.

This assertion requires a bit of explanation. Let us assume, for simplicity, that our domain Ω is bounded. Then we note that $A^2(\Omega)$ is a vector space over the field of complex numbers; and it is an infinite dimensional vector space because all holomorphic polynomials are in $A^2(\Omega)$. It can be checked that $A^2(\Omega)$, for Ω bounded, is separable (it is, after all, a subspace of the separable space $L^2(\Omega)$).

The inner product on A^2, as already noted, is

$$\langle f, g \rangle = \iint_{\Omega} f(z)\overline{g(z)} \, dA(z).$$

A collection of elements $\{\phi_\alpha\}_{\alpha \in A}$ is called ortho*normal* if each ϕ_α satisfies $\|\phi_\alpha\|^2 = \langle \phi_\alpha, \phi_\alpha \rangle = 1$ for all $\alpha \in A$ and $\langle \phi_\alpha, \phi_\beta \rangle = 0$ whenever $\alpha \neq \beta$. The collection $\{\phi_\alpha\}_{\alpha \in A}$ is termed *complete* if, whenever $\langle x, \phi_\alpha \rangle = 0$ for all α, then $x = 0$.

It is an exercise with Zorn's lemma to see that any Hilbert space, and A^2 in particular, has a complete orthonormal system. Because A^2 is separable, one can see that such a system must be countable. It is a fact from Hilbert space theory that if $\{\phi_j\}$ is a complete orthonormal system for a separable Hilbert space $\mathcal{H}$ and if $f \in \mathcal{H}$, then $f = \sum_j a_j \phi_j$, for some constants a_j, with convergence in the Hilbert space norm.

In general, it can be rather difficult to actually write down a complete orthonormal system for $A^2(\Omega)$. Fortunately, the unit disc $D \subseteq \mathbb{C}$ has enough symmetry that we can actually pull this off.

It is not difficult to see that $\{z^j\}_{j=0}^{\infty}$ is an orthogonal system for $A^2(D)$. That is, the elements are pairwise orthogonal, but they are not normalized to have unit length. We may confirm the first of these assertions by noting that if $j \neq k$, then

$$
\begin{aligned}
\langle z^j, z^k \rangle &= \iint_D z^j \overline{z^k} \, dA(z) \\
&= \int_0^1 \int_0^{2\pi} r^j e^{ij\theta} r^k e^{-ik\theta} \, d\theta r \, dr \\
&= \int_0^1 r^{j+k+1} \, dr \int_0^{2\pi} e^{i(j-k)\theta} \, d\theta \\
&= 0.
\end{aligned}
$$

The system $\{z^j\}$ is complete: If $\langle f, z^j \rangle = 0$ for every j, then f will have a null power series expansion and hence be identically zero. It remains to normalize these monomials so that we have a complete orthonormal system.

We calculate that

$$
\begin{aligned}
\iint_D |z^j|^2 \, dA(z) &= \int_0^1 \int_0^{2\pi} r^{2j} \, d\theta r \, dr \\
&= 2\pi \int_0^1 r^{2j+1} \, dr \\
&= \pi \cdot \frac{1}{j+1}.
\end{aligned}
$$

We conclude that

$$
\|z^j\| = \frac{\sqrt{\pi}}{\sqrt{j+1}}.
$$

Therefore the elements of our orthonormal system are

$$
\phi_j(z) = \frac{\sqrt{j+1} \cdot z^j}{\sqrt{\pi}}.
$$

Now, according to formula $(\star)$, the Bergman kernel is given by

$$K(z, \zeta) = \sum_{j=0}^{\infty} \phi_j(z) \cdot \overline{\phi_j(\zeta)}$$

$$= \sum_{j=0}^{\infty} \frac{(j+1)z^j \overline{\zeta^j}}{\pi}$$

$$= \frac{1}{\pi} \cdot \sum_{j=0}^{\infty} (j+1) \cdot (z\overline{\zeta})^j.$$

Observe that, in this instance, the convergence of the series is manifest (for both $|z| < 1$ and $|\zeta| < 1$). The convergence of the general orthonormal expansion in Lemma 1 is tricky, and we omit that proof.

Of course we can easily sum $\sum_j (j+1)\alpha^j$ by noticing that

$$\sum_{j=0}^{\infty} (j+1)\alpha^j = \frac{d}{d\alpha} \sum_{j=0}^{\infty} \alpha^{j+1}$$

$$= \frac{d}{d\alpha} \left[\alpha \cdot \frac{1}{1-\alpha} \right]$$

$$= \frac{1}{(1-\alpha)^2}.$$

Applying this result to our expression for $K(z, \zeta)$ yields that

$$K(z, \zeta) = \frac{1}{\pi} \cdot \frac{1}{(1 - z \cdot \overline{\zeta})^2}.$$

This is consistent with the formula which we obtained by conformal invariance in Subsection 3.1 for the Bergman kernel of the disc.

3.3 The Bergman Kernel by Way of Differential Equations

It is actually possible to obtain the Bergman kernel of a domain in the plane from the Green's function for that domain (see [EPS]). Let us now summarize the key ideas. Unlike the first two Bergman kernel constructions, the present one will work for *any* domain with C^2 boundary.

First, the fundamental solution for the Laplacian in the plane is the function

$$\Gamma(\zeta, z) = \frac{1}{2\pi} \log |\zeta - z|.$$

This means that $\Delta_\zeta \Gamma(\zeta, z) = \delta_z$. [Observe that δ_z denotes the Dirac "delta mass" at z and Δ_ζ is the Laplacian in the ζ variable.] Here the derivatives are interpreted in the sense of distributions. In more prosaic terms, the condition is that

$$\int \Gamma(\zeta, z) \cdot \Delta \phi(\zeta) \, d\tau \, d\eta = \phi(z)$$

for any C^2 function ϕ with compact support. We write, as usual, $\zeta = \tau + i\eta$. The reference [KR1, p. 35] provides details of this assertion.

Given a domain $\Omega \subseteq \mathbb{C}$, the *Green's function* is posited to be a function $G(\zeta, z)$ which satisfies

$$G(\zeta, z) = \Gamma(\zeta, z) + H(\zeta, z),$$

where $H(\zeta, z)$ is a particular harmonic function in the ζ variable. Moreover, it is mandated that $G(\cdot, z)$ vanishes on the boundary of Ω. One constructs the function H, for each fixed z, by solving a suitable Dirichlet problem. Again, the reference [KR1, p. 40, ff.] has all the details. Now we have

Proposition 2. *Let $\Omega \subseteq \mathbb{C}$ be a bounded domain with C^2 boundary. Let $G(\zeta, z)$ be the Green's function for Ω and let $K(z, \zeta)$ be the Bergman kernel for Ω. Then*

$$K(z, \zeta) = 4 \cdot \overline{\frac{\partial^2}{\partial \zeta \, \partial \overline{z}} G(\zeta, z)}. \tag{$\star\star$}$$

Proof. Our proof will use a version of Stokes's theorem written in the notation of complex variables. It says that if $u \in C^1(\overline{\Omega})$, then

$$\oint_{\partial U} u(\zeta)\, d\zeta = 2i \cdot \iint_U \frac{\partial u}{\partial \overline{\zeta}}\, d\tau \, d\eta, \qquad (**)$$

where again $\zeta = \tau + i\eta$. The reader is invited to convert this formula to an expression in τ and η and to confirm that the result coincides with the standard real-variable version of Stokes's theorem which can be found in any calculus book (see, e.g., [THO]).

Now we already know that

$$G(\zeta, z) = \frac{1}{4\pi} \log(\zeta - z) + \frac{1}{4\pi} \log \overline{(\zeta - z)} + H(\zeta, z). \qquad (\dagger)$$

Here we think of the logarithm as a multivalued holomorphic function; after we take a derivative, the ambiguity (which comes from an additive multiple of $2\pi i$) goes away.

Differentiating with respect to z (and using subscripts to denote derivatives), we find that

$$G_z(\zeta, z) = \frac{1}{4\pi} \frac{-1}{\zeta - z} + H_z(\zeta, z).$$

We may rearrange this formula to read

$$\frac{1}{\zeta - z} = -4\pi \cdot G_z(\zeta, z) + 4\pi H_z(\zeta, z).$$

We know that G, as a function of ζ, vanishes on $\partial\Omega$. Hence so does G_z. Let $f \in C^2(\overline{\Omega})$ be holomorphic on Ω. It follows that the Cauchy formula

$$f(z) = \frac{1}{2\pi i} \oint_{\partial\Omega} \frac{f(\zeta)}{\zeta - z}\, d\zeta$$

can be rewritten as

$$f(z) = \frac{2}{i} \oint_{\partial\Omega} f(\zeta) H_z(\zeta, z)\, d\zeta.$$

or

$$-2i \oint_{\partial \Omega} f(\zeta) H_z(\zeta, z) \, d\zeta = f(z).$$

Now we apply Stokes's theorem (in the complex form) to rewrite this last as

$$f(z) = 4 \cdot \iint_{\Omega} (f(\zeta) H_z)_{\bar{\zeta}} \, d\tau \, d\eta.$$

Since f is holomorphic and H is real-valued, we may conveniently write this last formula as

$$f(z) = 4 \cdot \iint_{\Omega} f(\zeta) \overline{H_{\zeta \bar{z}}} \, d\tau \, d\eta.$$

Now formula (†) tells us that $H_{\zeta \bar{z}} = G_{\zeta \bar{z}}$. Therefore we have

$$f(z) = \iint_{\Omega} f(\zeta) 4 \overline{G_{\zeta \bar{z}}} \, d\tau \, d\eta. \qquad (\ddagger)$$

With a suitable limiting argument, we may extend this formula from functions f which are $C^2(\overline{\Omega})$ to functions in $A^2(\Omega)$.

It is straightforward now to verify that $4 \overline{G_{\zeta \bar{z}}}$ satisfies the first three characterizing properties of the Bergman kernel—just by examining our construction. The crucial reproducing property is of course formula (‡). Then it follows that

$$K(z, \zeta) = 4 \cdot \overline{\frac{\partial^2}{\partial \zeta \partial \bar{z}} G(\zeta, z)}.$$

That is the desired result.   ∎

It is worth noting that the theorem we have just established gives a practical method to confirm the existence of the Bergman kernel—by relating it to the Green's function, whose existence is elementary.

Now let us calculate. Of course the Green's function of the unit disc D is

$$G(\zeta, z) = \frac{1}{2\pi} \log |\zeta - z| - \frac{1}{2\pi} \log |1 - \zeta \bar{z}|,$$

as a glance at any classical complex analysis text will tell us (see, for example, [COH] or [HIL]).

With formula ($\star\star$) in mind, we can make life a bit easier by writing

$$G(\zeta, z) = \frac{1}{4\pi} \log(\zeta - z) + \frac{1}{4\pi} \log(\overline{\zeta - z})$$
$$- \frac{1}{4\pi} \log(1 - \zeta\bar{z}) - \frac{1}{4\pi} \log\left(\overline{1 - \zeta\bar{z}}\right).$$

Here we think of the expression on the right as the concatenation of four multi-valued functions, in view of the ambiguity of the logarithm function. This ambiguity is irrelevant for us because the derivative of the Green's function is still well defined (i.e., the derivative annihilates additive constants).

Now we readily calculate that

$$\frac{\partial G}{\partial \bar{z}} = \frac{1}{4\pi} \cdot \frac{-1}{\overline{\zeta - z}} + \frac{1}{4\pi} \cdot \frac{\zeta}{1 - \zeta\bar{z}}$$

and

$$\frac{\partial^2 G}{\partial \zeta \partial \bar{z}} = \frac{1}{4\pi} \cdot \frac{1}{(1 - \zeta\bar{z})^2}.$$

In conclusion, we may apply Proposition 2 to see that

$$K(z, \zeta) = \frac{1}{\pi} \cdot \frac{1}{(1 - z \cdot \bar{\zeta})^2}.$$

This result is consistent with that obtained in the first two calculations (Subsections 4.3.1, 4.3.2).

4. About the Bergman Metric

The Bergman kernel is an important conformal invariant. For the purposes of the present book, this invariance manifests itself most strikingly in the form of the Bergman (sometimes called the Poincaré-Bergman) metric.

In fact we first notice that when Ω is a bounded domain, $K(z, z)$ is always positive. This assertion is most easily seen from the series construction of the Bergman kernel. At any point z_0 we have

$$0 = K(z_0, z_0) = \sum_{j=1}^{\infty} |\phi_j(z_0)|^2.$$

It is immediate that K is non-negative on the diagonal. If there were a z_0 such that $K(z_0, z_0) = 0$, then the last sum would be zero. It would then follow that $\phi_j(z_0) = 0$ for all j. But, because the $\{\phi_j\}$ are a complete orthonormal system for A^2, and any $f \in A^2$ can be expanded in terms of the ϕ_j, it would then follow that $f(z_0) = 0$ for all $f \in A^2$. This is false for the function $f \equiv 1$.

Since $K(z, z)$ is always positive, we may consider the quantity $\log K(z, z)$. We define the *Bergman metric* to be that metric given by the weight function

$$\rho_\Omega(z) = \rho(z) = \sqrt{\frac{\partial^2}{\partial z \partial \overline{z}} \log K_\Omega(z, z)}.$$

Notice that the logarithm under the square root is certainly real, and the double derivative is real since it does not change under conjugation. That the expression under the square root sign is positive follows because we may differentiate the series expansion $\sum_j \phi_j \overline{\phi}_j$ of K:

$$\frac{\partial^2}{\partial z \partial \overline{z}} \log K(z, z) = \frac{\partial}{\partial z} \left[\left(\frac{1}{\sum_j \phi_j \overline{\phi}_j} \right) \cdot \sum_j \phi_j \overline{\phi}_j' \right]$$

$$= -\frac{1}{\left(\sum_j \phi_j \overline{\phi}_j\right)^2} \cdot \sum_j \phi_j' \overline{\phi}_j \cdot \sum_j \phi_j \overline{\phi}_j'$$

$$+ \frac{1}{\left(\sum_j \phi_j \overline{\phi}_j\right)} \cdot \sum_j \phi_j' \overline{\phi}_j'$$

$$= \frac{1}{\sum_j |\phi_j|^2} \left[\sum_j |\phi_j'|^2 - \frac{1}{\sum_j |\phi_j|^2} \cdot \left| \sum_j \phi_j \overline{\phi}_j' \right|^2 \right].$$

Of course the expression in square brackets is nonnegative by the Cauchy-Schwarz inequality.

Our first claim is that the Bergman metric ρ_Ω is in fact an invariant metric:

Proposition 1. *Let Ω_1, Ω_2 be bounded, conformally equivalent domains. Let $\Phi : \Omega_1 \to \Omega_2$ be a conformal mapping. Then*

$$\Phi^* \rho_{\Omega_2}(z) = \rho_{\Omega_1}(z).$$

Proof. We calculate that

$$\Phi^* \rho_{\Omega_2}(w) = |\Phi'(w)| \cdot \rho_{\Omega_2}(\Phi(w))$$

$$= \sqrt{\Phi'(w) \left[\frac{\partial^2}{\partial z \partial \overline{z}} \log K_{\Omega_2}(z, z) \right] \Bigg|_{z = \Phi(w)} \overline{\Phi'(w)}}$$

$$= \sqrt{\frac{\partial^2}{\partial w \partial \overline{w}} \log \left[\Phi'(w) \cdot K_{\Omega_2}(\Phi(w), \Phi(w)) \overline{\Phi'(w)} \right]}$$

Here we have inserted terms with derivative zero, and have also applied the chain rule. Now we have that this expression

$$= \sqrt{\frac{\partial^2}{\partial w \partial \overline{w}} \log K_{\Omega_1}(w, w)}$$

$$= \rho_{\Omega_1}(w).$$

Thus the Bergman metric is conformally invariant. ∎

Clearly any metric that is invariant under conformal mappings is a matter of great interest. This book provides ample proof of these assertions. Let us next calculate the Bergman metric for the disc D.

Example 1. We know that the Bergman kernel for the disc D is

$$K(z, \zeta) = \frac{1}{\pi} \cdot \frac{1}{(1 - z \cdot \overline{\zeta})^2}.$$

Therefore

$$K(z, z) = \frac{1}{\pi} \cdot \frac{1}{(1 - |z|^2)^2}.$$

In particular, we see (as we already know *a priori*) that K restricted to the diagonal is positive. Thus $\log K(z, z)$ makes sense and

$$\log K(z, z) = -\log \pi - 2 \log(1 - |z|^2).$$

As a result,

$$\frac{\partial}{\partial \bar{z}} \log K(z, z) = \frac{-2}{1 - |z|^2} \cdot (-z) = \frac{2z}{1 - |z|^2}.$$

Therefore

$$\frac{\partial^2}{\partial z \partial \bar{z}} \log K(z, z) = \frac{2}{(1 - |z|^2)^2}.$$

In conclusion,

$$\rho_D(z) = \frac{\sqrt{2}}{1 - |z|^2}.$$

This is just the same (up to a constant multiple) as the Poincaré metric which we calculated for the disc in Section 1.4. ∎

In fact it can be argued—using ideas from Section 1.4—that the only invariant metric (up to a constant multiple) on the disc—of the type we have been studying in this book—is the Poincaré metric. So it is no surprise that the Poincaré and Bergman metrics coincide. By contrast, the reader may check for himself that the metric

$$\psi(z, w) = \log \left| \frac{z - w}{1 - z \cdot \bar{w}} \right|$$

is an invariant metric on the disc (i.e., conformal transformations ϕ of the disc preserve distances in this metric, so that $\psi(\phi(z), \phi(w)) =$

$\psi(z, w)$) but that ψ is *not* of the form studied in the present book. There is no contradiction because ψ is *not obtained by integration against a weight function ρ*. [The metric defined here is sometimes termed the "pseudohyperbolic metric."]

5. More on the Bergman Metric

The fact that the Bergman metric is obtained as the Laplacian of a potential function (in this case $\log K(z, z)$) means that the metric is a *Kähler* metric. This means, roughly speaking, that the geometric structure and the complex structure are compatible in a certain sense—see [KOK] for the full story about Kähler metrics. The potential function is automatically real analytic, hence so is the metric itself. These are powerful facts when the Bergman metric is used in geometric calculations.

We know from classical function theory (see, for instance, [GRK] and the discussion of Carathéodory's theorem in Chapter 0) that a conformal mapping of two bounded domains, each simply connected and having boundary consisting of a single Jordan curve, will extend univalently and continuously to the boundary. So the conformal mapping becomes a homeomorphism of the closures of the domains. This fact is useful in transferring the function theory from one domain to the other. It is useful in more advanced contexts to know that, when two domains have smooth boundaries and are conformally equivalent, then a conformal mapping of the domains will extend smoothly to the boundaries. This problem was first explored by Painlevé in his Paris thesis [PAI] in 1888. He proved, in particular, that if each domain has C^∞ boundary then the conformal mapping and its inverse each extend in a C^∞ manner to the boundary. Painlevé's result was generalized and refined by Kellogg and Warschawski, among others.

In modern times, the Bergman kernel and metric have been used to simplify and clarify (and also make more geometrical) the study of boundary smoothness of conformal mappings. This idea was first explored by C. Fefferman in [FEF1] in a much broader setting. It was later simplified by S. Bell and his collaborators. A complete exposition, in the context of one complex variable, appears in [BEK] or [GRK].

6. Application to Conformal Mapping

A classical result from complex function theory is this:

Theorem 1. *Let $A = \{z \in \mathbb{C} : 1 < |z| < R\}$. If ϕ is a conformal mapping of A to itself, then ϕ is either a rotation or an inversion of the form $z \mapsto R/z$.*

In this section we would like to conduct a mathematically precise but somewhat informal discussion, from the point of view of Bergman geometry, of why this theorem is true. What is rewarding about this treatment is that it illustrates how a geometer thinks. Moreover, all of this purely geometric reasoning can, with some effort, be made absolutely rigorous and analytical. We are pleased to thank Robert E. Greene and John McCarthy for helpful discussions of this result.

It is interesting to note that it is quite difficult to compute the Bergman kernel and metric for an annulus. One can certainly see (using Laurent series) that the monomials $\{z^j\}_{j=-\infty}^{\infty}$ form an orthogonal system on an annulus centered at the origin. And one can use a little calculus to normalize these to an orthonormal system. But actually performing the necessary summation is virtually intractable (and involves elliptic functions [BER, pp. 9–10]). Fortunately, the proof which we are about to present requires no detailed knowledge of the Bergman metric of the annulus. Indeed it uses only the fact that the metric blows up at the boundary. We will get this information for free from the following elegant and surprising result of K. T. Hahn:

Proposition 2. *Let $\Omega \subseteq \mathbb{C}$ be a bounded domain. Let ρ_Ω denote the Bergman metric on Ω and let F_C^Ω be the Carathéodory metric. Then*

$$\rho_\Omega(z) \geq F_C^\Omega(z)$$

for every $z \in \Omega$.

Of course we already know that the Carathéodory metric F_C^Ω is complete, hence blows up at the boundary—see Theorem 6 of Section 3.3. Proposition 2 (see [HAH] for the proof) tells us immediately that the Bergman metric ρ is complete, hence blows up at the boundary of Ω. See [APF] for more on the boundary behavior of the Bergman kernel and its derivatives.

We use the Bergman metric to prove our results in this section, rather than the Carathéodory or Kobayashi metrics, because we will be doing calculations with geodesics. So we need a smooth, complete metric (details on this point follow). We will also utilize the real analyticity of the Bergman kernel and metric in an interesting and surprising way.

We will use the language of geodesics. Going by the book, a geodesic is defined by a differential equation. For our purposes here, we may think of a geodesic as a locally length-minimizing curve. Now fix an annulus $\mathcal{A} = \{z \in \mathbb{C} : 1 < |z| < R\}$. We will be using standard polar coordinates (r, θ). At any point $p \in \mathcal{A}$, a vector in the tangent space decomposes into a component in the $\partial/\partial r$ direction and a component in the $\partial/\partial \theta$ direction. One way to look at the metric is that it assigns a length to $\partial/\partial r$ and to $\partial/\partial \theta$ at each point of the annulus. Consider the set M of points where the Bergman length of $\partial/\partial \theta$ is minimal. Such points exist just because the Bergman metric blows up at the boundary of the domain. What geometric properties will the set M have?

First, the set must be rotationally invariant—because the metric will be rotationally invariant (i.e., the rotations are conformal self-maps of $\mathcal{A}$). Thus M is a union of circles centered at the origin. And M is certainly a closed set by the continuity of the metric. The set has no interior because the Bergman metric ρ is given by a real analytic function (it is the second derivative of the logarithm of the real analytic Bergman kernel), and the zero set of a non-trivial real analytic function can have no interior—see [KRP1]. In fact we may also note that $\|\partial/\partial \theta\|_\rho$ is real analytic in the radial r direction, hence (by the same reasoning) we claim that M can have only finitely many circles in it. More precisely, let $g(r, \theta) = \|\partial/\partial \theta\|_{\rho,(r,\theta)}$. We have already noted that this function is independent of θ, so we may consider $g(r) = \|\partial/\partial\theta\|_{\rho,r}$. This function will assume its minimum value at points r where $g'(r) = 0$. Since the metric is complete, we know that the set of such points forms a compact subset of the interval $(0, R)$. If the set is infinite, then it has an accumulation point and, therefore, the real analytic function g' is identically zero. That is impossible, again by the completeness of the metric. Therefore the set M is finite.

Now it is easy to see that any curve (circle) in M will also be one of the curves (the curves that go once around the hole in the middle of the annulus) that minimizes arc length in the Bergman metric; and vice versa. This is so because we have already selected the curve to have $\partial/\partial\theta$ length as small as possible; a curve whose tangents have components in the $\partial/\partial r$ direction will *a fortiori* be longer. In more detail, the length of any curve $\gamma(t), 0 \leq t \leq 1$ is calculated by

$$\ell_\rho(\gamma) = \int_0^1 \|\gamma'(t)\|_{\rho,\gamma(t)}\, dt = \int_0^1 \|\gamma_r'(t) + \gamma_\theta'(t)\|_{\rho,\gamma(t)}\, dt,$$

where γ_r' and γ_θ' are, respectively, the normal and tangential components of γ'. Clearly, if we construct a new curve $\tilde{\gamma}$ by integrating the vector field γ_θ', then the result is a curve that is shorter than γ.

Thus any circle in M is a length-minimizing geodesic. And the converse is true as well.

Now let ϕ be a conformal self-map of $\mathcal{A}$. Then, as a result of the considerations in the last paragraph, ϕ will map circles in M (concentric with the annulus) to circles in M (concentric with the annulus). But more is true: *any* circle c that is centered at 0 has constant Bergman distance from any one given circle C in M. Let τ_c be the Bergman distance from the arbitrary circle c to the fixed circle C. Then c will be mapped by ϕ to another circle in $\mathcal{A}$ that has Bergman distance τ_c from $\phi(C)$. In short, ϕ maps circles to circles. And the same remark applies to ϕ^{-1}.

Now the orthogonal trajectories to the family of circles centered at P will of course be the radii of the annulus $\mathcal{A}$. Therefore (by conformality) these radii will get mapped to radii. Let $\mathcal{R}$ be the intersection of the positive real axis with the annulus $\mathcal{A}$. After composition with a rotation, we may assume that ϕ maps $\mathcal{R}$ to $\mathcal{R}$.

If M has just one circle in it, then that circle C must be fixed. Since each circle c in the annulus (centered at P) has a distance τ_c from C, then its image under ϕ will have the same distance from C. Thus either c is fixed or it is sent to its image under inversion. By continuity, whatever choice is valid for c will be valid for all other circles. Thus, in this case, the map ϕ is either the identity or inversion.

Now suppose that M contains at least two circles. The only possibilities for the action of ϕ on $M \cap \mathcal{R}$ are preservation of the order of the points or inversion of the order (as any other permutation of the points is ruled out by elementary topology of a conformal mapping). By composing ϕ with an inversion ($z \mapsto R/z$), we may assume that ϕ does *not* act as an inversion on $\mathcal{R}$. But of course ϕ must preserve $M \cap \mathcal{R}$. We conclude that ϕ must be the identity.

What we have just proved is that, after we normalize ϕ so that it maps the positive real axis to itself, then in fact ϕ must be the identity (or an inversion). In other words, the original map ϕ must be a rotation (or an inversion). ∎

CHAPTER 5

A Glimpse of Several Complex Variables

0. Functions of Several Complex Variables

At a naive level, the analysis of several real variables is much like the real analysis of one variable; the main difference is that one deals with n-tuples of reals instead of scalars and one needs matrices to keep track of information. Of course deeper study reveals much complexity and richness in analysis of several real variables. It is noteworthy that, for the most part, this richness was discovered rather late in the history of the subject—mostly in the last fifty years.

The history of analysis of several complex variables is quite different. Early in the subject, in the first decade of the twentieth century, two remarkable discoveries indicated that this area has an incredible depth and variety which one complex variable does not even hint at. Even today we have only scratched the surface of several complex variables.

Let us briefly discuss the two developments which established several complex variables as a subject in its own right. The first is related to the Riemann mapping theorem. As we have discussed throughout the present volume, Riemann's theorem asserts that, with the single exception of the plane, any domain topologically equivalent to the disc is conformally equivalent to it. One might expect an analogue to this result in two complex variables. But, to begin with, what is the analogue of the disc in $\mathbb{C}^2$? Two candidates come to mind: the ball

$$\{(z_1, z_2) \in \mathbb{C}^2 : |z_1|^2 + |z_1|^2 < 1\}$$

and the bidisc

$$\{(z_1, z_2) \in \mathbb{C}^2 : |z_1| < 1, \ |z_2| < 1\}.$$

Which of these should serve as the model domain for the multivariable Riemann mapping theorem? Before answering that question, perhaps one should ask whether there is a biholomorphic equivalence between the ball and the bidisc. [According to a result of Liouville (see [DFN]), there are none but trivial conformal mappings in dimensions three and higher. As a result, in several complex variables we consider *biholomorphic* mappings. These are mappings which are holomorphic, one-to-one, and onto. The holomorphicity of the inverse is automatic. See further discussion below in Section 1.] If there is such an equivalence, then the other question need not be answered; if there is no equivalence, then perhaps there is no Riemann mapping theorem.

As Poincaré discovered, there is in fact no equivalence between the ball and the bidisc. In the last two decades, it has been discovered that, generically, two domains in $\mathbb{C}^2$ are not biholomorphically equivalent. Thus the entire question of the Riemann mapping theorem turns into a rather complex subject in itself—that of classifying domains according to biholomorphic equivalence. One of the main things that we shall do in this chapter is to formulate carefully and to prove Poincaré's theorem concerning the inequivalence of the ball and the bidisc.

The other big development which occurred near the turn of the twentieth century is Hartogs's theorem about analytic continuation. Although we cannot treat this topic in detail here, we should like to discuss it briefly in order to further convey the flavor of several complex variables. So that we may keep matters as simple as possible, we shall avoid formal definitions in this discussion.

A domain U in complex space is called a *domain of holomorphy* if there is a holomorphic function defined on U which cannot be analytically continued to any open set which properly contains U. It turns out (see [KR1]) that the Weierstrass theorem of classical function theory can be used to show that *any* open set U in the plane is a domain of holomorphy. Briefly, let U be such an open set. Let $S \subseteq U$ be a set in U such that (i) S has no accumulation point in the interior of U and (ii) S

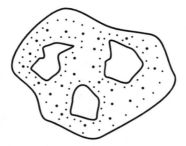

Figure 1.

accumulates at every boundary point of U—see Figure 1. Then Weierstrass's theorem specifies the existence of a holomorphic function f on U which vanishes precisely on S (and nowhere else). If this f were to analytically continue past any boundary point p of U, then p would be an interior accumulation point of the zero set of f. Thus f would be identically zero on some connected component of U, a contradiction. We conclude that f is a holomorphic function on U which cannot be continued to any larger open set. See [GRK, p. 270] for the details of this construction.

What Hartogs discovered is that, in dimensions 2 and higher, some domains are domains of holomorphy and others are not. For example, he proved that the domain

$$U = \{(z_1, z_2) : |z_1| <, 2, \ |z_2| < 2\}$$
$$\setminus \{(z_1, z_2) : |z_1| \le 1, \ |z_2| \le 1\}$$

has the property that any function holomorphic on U can be analytically continued to

$$\widehat{U} = \{(z_1, z_2) : |z_1| < 2, \ |z_2| < 2\}.$$

The full proof of this result may be found in [KR1]. As previously noted, this theorem has no analogue in one complex variable and begs the question: "Which domains in several complex variables have the property exhibited by Hartogs's phenomenon?" A necessary geometric condition on the boundary of U was discovered by E.E. Levi, and the problem of proving its sufficiency became known as the Levi problem. The Levi problem was solved in considerable generality in the 1940's and 1950's, but several related questions still remain open.

The point which we have tried to make in this brief essay is that many of the most interesting questions in several complex variables never come up in one complex variable, because they make no sense in that context. What we can hope to accomplish in this chapter is to give the reader a glance at the subject and suggest some of the techniques that can be used effectively. One of our accomplishments will be to give two fairly easy and elegant proofs of Poincaré's theorem.

1. Basic Concepts

We introduce here only the few definitions and pieces of notation which we shall need for our immediate purposes. For a more comprehensive introduction to several complex variables, see [KR1, KR2, KR4]. For simplicity of notation, we confine attention to *two* complex variables.

We consider domains $U \subseteq \mathbb{C}^2$, where

$$\mathbb{C}^2 \equiv \mathbb{C} \times \mathbb{C}$$

and (z_1, z_2) is a typical element. Recall that $\mathbb{C}^2$ may be identified with $\mathbb{R}^4$ by

$$\mathbb{C}^2 \ni (z_1, z_2) \longleftrightarrow (x_1 + iy_1, x_2 + iy_2)$$

$$\longleftrightarrow (x_1, y_1, x_2, y_2) \in \mathbb{R}^4.$$

Thus a function of two complex variables is continuous if it is continuous as a function of four real variables, it is continuously differentiable if it is continuously differentiable as a function of four real variables, and so forth.

Let $U \subseteq \mathbb{C}^2$ be a domain, that is, a connected open set. Let z and w be complex numbers. We define

$$U_w = \{z_1 \in \mathbb{C} : (z_1, w) \in U\}$$

and

$$U^z = \{z_2 \in \mathbb{C} : (z, z_2) \in U\}.$$

Definition 1. A continuously differentiable function f on a domain U is said to be *holomorphic* if, whenever $z \in \mathbb{C}$ satisfies $U^z \neq \emptyset$, then $f(z, \cdot)$ is holomorphic on U^z (as a function of *one* complex variable) and whenever $w \in \mathbb{C}$ satisfies $U_w \neq \emptyset$, then $f(\cdot, w)$ is holomorphic on U_w (as a function of *one* complex variable).

Example 1. The function

$$f(z_1, z_2) = \frac{z_1}{z_2 + 1}$$

is holomorphic on the domain

$$U_1 = \{(z_1, z_2) : z_2 \neq -1\}.$$

The function

$$g(z_1, z_2) = z_1(\bar{z}_2)^2$$

is not holomorphic on any domain. The function

$$h(z_1, z_2) = \sum_j (z_1)^j (z_2)^j$$

is holomorphic on

$$U_2 = \{(z_1, z_2) : |z_1 z_2| < 1\}. \qquad \blacksquare$$

In several complex variables there are two natural notions of "neighborhood": the ball and the bidisc. If $P = (P_1, P_2) \in \mathbb{C}^2$ and $r > 0$, then we let

$$\mathcal{B}(P, r) = \{z \in \mathbb{C}^2 : |z_1 - P_1|^2 + |z_2 - P_2|^2 < r^2\}$$

and

$$D^2(P, r) = \{z \in \mathbb{C}^2 : |z_1 - P_1| < r, \ |z_2 - P_2| < r\}$$

be, respectively, the open ball and open bidisc of center P and radius r. We also let $\overline{\mathcal{B}}(P,r)$ and $\overline{D}^2(P,r)$ denote, respectively, the closed ball and closed bidisc (defined by replacing the symbol $<$ with the symbol $\leq$).

A basic tool in complex analysis is the Cauchy integral formula and its variants. We now derive one such.

Proposition 2. *If f is holomorphic on a neighborhood of $\overline{D}^2(P,r)$ (the closure of the bidisc), then for all $(z_1, z_2) \in D^2(P,r)$ we have*

$$f(z_1, z_2) = \frac{1}{(2\pi i)^2} \oint_{\partial D(P_1, r)} \oint_{\partial D(P_2, r)} \frac{f(\zeta_1, \zeta_2)}{(\zeta_1 - z_1)(\zeta_2 - z_2)} \, d\zeta_2 d\zeta_1.$$

Proof. The function $f(\cdot, z_2)$ is holomorphic in a neighborhood of $\overline{D}(P_1, r)$. Therefore the one-variable Cauchy integral formula yields, for each fixed value of z_2,

$$f(z_1, z_2) = \frac{1}{2\pi i} \oint_{\partial D(P_1, r)} \frac{f(\zeta_1, z_2)}{(\zeta_1 - z_1)} \, d\zeta_1.$$

Now apply the one variable Cauchy integral formula to the function $f(\zeta_1, z_2)$ in the integrand *in the second variable:* the desired formula follows.   ∎

Corollary 2.1. *A holomorphic function of two complex variables is infinitely differentiable as a function of four real variables.*

Proof. Differentiate under the integral sign, just as in one complex variable.   ∎

Corollary 2.2. *A function f as in the Proposition has a power series expansion*

$$f(z_1, z_2) = \sum a_{jk}(z_1 - P_1)^j (z_2 - P_2)^k$$

which converges absolutely and uniformly on $\overline{D}^2(P, r)$. *The coefficients are given by*

$$a_{j,k} = \frac{1}{j!k!} \left(\frac{\partial}{\partial z_1} \right)^j \left(\frac{\partial}{\partial z_2} \right)^k f(P).$$

Proof. The proof follows familiar lines from one complex variable.
Since f is holomorphic on a neighborhood of $\overline{D}^2(P, r)$, we may select
$r' > r$ so that f is holomorphic on a neighborhood of $\overline{D}^2(P, r')$. We
write

$$f(z_1, z_2) = \frac{1}{(2\pi i)^2} \oint_{\partial D(P_1, r')} \oint_{\partial D(P_2, r')} \frac{f(\zeta_1, \zeta_2)}{(\zeta_1 - z_1) \cdot (\zeta_2 - z_2)} \, d\zeta_2 \, d\zeta_1.$$

$$(*)$$

However,

$$
\begin{aligned}
\frac{1}{(\zeta_1 - z_1)} &= \frac{1}{(\zeta_1 - P_1) - (z_1 - P_1)} \\
&= \frac{1}{(\zeta_1 - P_1)} \frac{1}{1 - ((z_1 - P_1)/(\zeta_1 - P_1))} \\
&= \sum_{j=0}^{\infty} \frac{(z_1 - P_1)^j}{(\zeta_1 - P_1)^{j+1}}.
\end{aligned}
\qquad (**)
$$

Similarly,

$$\frac{1}{(\zeta_2 - z_2)} = \sum_{j=0}^{\infty} \frac{(z_2 - P_2)^j}{(\zeta_2 - P_2)^{j+1}}. \qquad (***)$$

Convergence is absolute and uniform on $\{|z_1 - P_1| \le r\}$ and $\{|z_2 - P_2| \le r\}$ respectively. Thus we may substitute $(**)$ and $(***)$ into $(*)$
and switch the order of summation and integration to obtain

$$f(z_1, z_2) = \sum_{j=0}^{\infty} \sum_{k=0}^{\infty} \left[\frac{1}{(2\pi i)^2} \oint_{\partial D(P_1, r)} \oint_{\partial D(P_2, r)} \right.$$

$$\left. \frac{f(\zeta_1, \zeta_2)d\zeta_2 d\zeta_1}{(\zeta_1 - P_1)^{j+1}(\zeta_2 - P_2)^{k+1}} \right]$$

$$\times (z_1 - P_1)^j (z_2 - P_2)^k$$

$$\equiv \sum_{j,k=0}^{\infty} a_{j,k} \cdot (z_1 - P_1)^j (z_2 - P_2)^k.$$

Here

$$a_{j,k} = \frac{1}{(2\pi i)^2} \oint_{\partial D(P_1, r)} \oint_{\partial D(P_2, r)} \frac{f(\zeta_1, \zeta_2)d\zeta_2 d\zeta_1}{(\zeta_1 - P_1)^{j+1}(\zeta_2 - P_2)^{k+1}}.$$

The familiar Cauchy theory of one complex variable now tells us that

$$a_{j,k} = \frac{1}{j!k!} \left(\frac{\partial}{\partial z_1}\right)^j \left(\frac{\partial}{\partial z_2}\right)^k f(P). \qquad \blacksquare$$

For simplicity let us now consider power series

$$S \sim \sum a_{j,k} z_1^j z_2^k$$

expanded about $0 = (0, 0)$. Notice that if S converges absolutely at a point (z_1, z_2) then it also converges absolutely and uniformly at the points $(\mu_1 z_1, \mu_2 z_2)$ where $|\mu_1| \leq 1$, $|\mu_2| \leq 1$. This motivates the following definition.

Definition 3. A domain U is called *complete circular* if whenever $(z_1, z_2) \in U$ and $|\mu_1|, |\mu_2| \leq 1$ then $(\mu_1 z_1, \mu_2 z_2) \in U$. Note in passing that a complete circular domain automatically contains 0.

Proposition 4. *Let*

$$S \sim \sum a_{j,k} z_1^j z_2^k$$

be a power series. Then

$$\mathcal{C} = \{z = (z_1, z_2) : S \text{ converges absolutely in a neighborhood of } z\}$$

is an open, complete circular domain. We call C the domain of convergence of S.

Proof. Combine the definition of C with that of complete circular domain and the remark preceding it. ∎

Although we did not assert the converse of the Proposition, it nevertheless suggests that the ball and the bidisc might each be domains of convergence for some power series. Indeed,

$$S \sim \sum z_1^j z_2^k$$

converges precisely on $D^2(0, 1)$ and on no larger open set. The power series

$$T \sim \sum_j \left((z_1)^2 + (z_2)^2 \right)^j$$

(when rearranged as a sum of *monomials*) converges absolutely on the open ball, and on no larger open set. (Exercise: fill in the details of these assertions.)

One of the themes of this book has been to consider conformal self-maps of domains. It turns out that, as previously noted, "conformal" is the wrong notion for functions of several complex variables. The correct morphisms for our purposes are given in the following definition:

Definition 5. Let U_1, U_2 be domains in $\mathbb{C}^2$. A function

$$F : U_1 \longrightarrow U_2$$

is called a *biholomorphic mapping* if

$$F = (f_1(z_1, z_2), f_2(z_1, z_2))$$

is a pair of holomorphic functions, F is injective, F is surjective, and F^{-1} is holomorphic.

Remark. It turns out that the condition that F^{-1} be holomorphic is redundant; however this is hard to prove (see [KR1] for this result of Osgood) so, for simplicity, we state the condition explicitly.   ∎

Example 2. Let $U = D^2(0, 1)$. Then

$$f(z_1, z_2) = \left(\frac{z_2 - 1/2}{1 - (1/2)z_2}, \ i \cdot z_1 \right)$$

is a biholomorphic mapping of U to U.   ∎

As in the case of one variable, we let $\text{Aut}(U)$ denote the collection of biholomorphic self-maps of U.

Example 3. Let $U = \mathcal{B}(0, 1)$. Then

$$f(z_1, z_2) = \left(\frac{z_1 - 1/2}{1 - (1/2)z_1}, \ \frac{(\sqrt{3}/2)z_2}{1 - (1/2)z_1} \right)$$

is a biholomorphic mapping of the domain U. Of course any unitary rotation is also a biholomorphic mapping of U.

Exercise. If $U \subseteq \mathbb{C}^2$ is a domain then $\text{Aut}(U)$ forms a group under composition of mappings.   ∎

2. The Automorphism Groups of the Ball and Bidisc

We shall calculate the automorphism groups of $\mathcal{B}(0, 1)$ and $D^2(0, 1)$. We begin with some preliminary material.

Lemma 1. *If f_j are holomorphic functions on U and $f_j \to f$ uniformly on compact subsets of U, then any sequence of derivatives*

$$\left(\frac{\partial}{\partial z_1} \right)^{\ell} \left(\frac{\partial}{\partial z_2} \right)^{k} f_j$$

converges to

$$\left(\frac{\partial}{\partial z_1}\right)^\ell \left(\frac{\partial}{\partial z_2}\right)^k f,$$

uniformly on compact sets.

Proof. The argument is the same as in one complex variable (Corollary 5.1 of Section 0.1): Fix $P \in U$ and choose $r > 0$ such that $\overline{D}^2(P, r) \subseteq U$. Express f_j on $D^2(P, r)$ as a Cauchy integral of f_j on $\partial D^2(P, r)$. The result is then immediate from differentiation under the integral sign. ■

Lemma 2. *If f is holomorphic on a domain $U \subseteq \mathbb{C}^2$ and if f vanishes on an open subset $W \subseteq U$ then $f \equiv 0$ on U.*

Proof. Let

$$C = \{z \in U : f(z) = 0 \text{ in a neighborhood of } z\}.$$

Then C is clearly open and nonempty. On the other hand, if $z \in \partial C \cap U$, then f and all its derivatives vanish at z. Thus the power series expansion for f in a neighborhood of z is identically 0, and z cannot be in ∂C. It follows that C has no boundary points in U. Therefore it must hold that $C = U$. ■

Lemma 3. *Let $\mathcal{F}$ be a family of holomorphic functions on a domain $U \subseteq \mathbb{C}^2$ such that*

$$|f(z)| \le M < \infty, \qquad \forall f \in \mathcal{F}.$$

Then $\mathcal{F}$ is a normal family.

Proof. It is enough to see that $\mathcal{F}$ is normal on any closed bidisc $\overline{D}^2(P, r) \subseteq U$. Let $\overline{D}^2(P, r') \subseteq U$ be a slightly larger bidisc. Differentiate the Cauchy integral formula on $D^2(P, r')$; the standard estimates from one variable then give

$$\left|\frac{\partial f}{\partial z_j}(z)\right| \le \frac{M}{r'-r}, \qquad \forall z \in \overline{D}^2(P,r), \ f \in \mathcal{F}, \quad j = 1, 2.$$

It follows that $\mathcal{F}$ is an equicontinuous family on $\overline{D}^2(P,r)$. Therefore the result follows from the Ascoli/Arzelà theorem. ■

Definition 4. If U_1, U_2 are domains in $\mathbb{C}^2$ and

$$f : U_1 \longrightarrow U_2$$

is holomorphic then define $\mathrm{Jac}_{\mathbb{C}} f(z), z \in U_1$, (the *Jacobian matrix* of f at z) to be the matrix

$$\begin{pmatrix} \dfrac{\partial f_1}{\partial z_1}(z) & \dfrac{\partial f_1}{\partial z_2}(z) \\[2ex] \dfrac{\partial f_2}{\partial z_1}(z) & \dfrac{\partial f_2}{\partial z_2}(z) \end{pmatrix}.$$

Remark. Notice that $\mathrm{Jac}_{\mathbb{C}} f(z)$ is distinct from the real Jacobian, $\mathrm{Jac}_{\mathbb{R}} f$, of calculus: the latter would be a 4×4 matrix which arises from treating f as a function from a domain in $\mathbb{R}^4$ to a domain in $\mathbb{R}^4$. ■

Next we prove a Schwarz lemma in two complex variables. Although it could be proved in an *ad hoc* manner for the ball and the bidisc, the proof in general is not essentially more difficult; it also allows us to recall a nice argument which we saw earlier in Section 3.2.

Proposition 5. (**H. Cartan**). *Let $U \subseteq \mathbb{C}^2$ be a bounded domain and $P \in U$. Let $f : U \to U$ be holomorphic and assume that $f(P) = P$. If*

$$\mathrm{Jac}_{\mathbb{C}} f(P) = \mathrm{id},$$

then

$$f(z) \equiv z.$$

Proof. Assume without loss of generality that $P = 0$. Seeking a contradiction, we assume that the conclusion is false.

In a neighborhood of $0 \in U$ we may expand f in a power series. The hypotheses about f imply that this power series has the form

$$f(z_1, z_2) = 0 + z + A_m(z) + \cdots$$

where $A_m(z)$ represents the first nonvanishing monomial after z, A_m having degree $m \geq 2$. Observe that $f = (f_1, f_2)$ is an ordered pair and that each summand on the right-hand side of our equation is an ordered pair. Define

$$f^1(z_1, z_2) = f$$
$$f^2(z_1, z_2) = f \circ f$$
$$\cdots$$
$$f^j(z_1, z_2) = f^{j-1} \circ f, \qquad \forall j \geq 2.$$

Since U is bounded, the family $\{f^j\}$ forms a normal family (by Lemma 3). Therefore there is a subsequence f^{j_ℓ} converging to a limit function $\widetilde{f}$. By Lemma 1, the mth derivatives of f^{j_ℓ} at P also converge to the corresponding derivatives of $\widetilde{f}$ at P.

On the other hand, direct calculation shows that

$$f^2 = z + 2A_m + \cdots$$
$$f^3 = z + 3A_m + \cdots$$
$$\text{etc.}$$

and we see that in fact the mth derivatives of f^j at $P = 0$ blow up. The only way to resolve this contradiction is to conclude that $A_m \equiv 0$ near P. Therefore $A_m \equiv 0$ on U. We conclude that $f(z) \equiv z$, as desired.

∎

Proposition 6. *Let U be a bounded complete circular domain. If*

$$f : U \longrightarrow U$$

is a biholomorphic mapping and $f(0) = 0$ then f is linear.

Remark. This is a generalization to two variables of the fact that a conformal map of the disc to the disc that preserves the origin is in fact a rotation. ∎

Proof of the Proposition. Let $\theta \in [0, 2\pi)$ and

$$\rho_\theta(z_1, z_2) = (e^{i\theta} z_1, e^{i\theta} z_2).$$

Consider the map

$$g = \rho_{-\theta} \circ f^{-1} \circ \rho_\theta \circ f.$$

Then

$$\mathrm{Jac}_{\mathbb{C}} g(0) = \rho_{-\theta} \circ (\mathrm{Jac}_{\mathbb{C}} f(0))^{-1} \circ \rho_\theta \circ (\mathrm{Jac}_{\mathbb{C}} f(0)).$$

Here we think of $\rho_{\pm\theta}$ as a linear map given by the matrix

$$\begin{pmatrix} e^{\pm i\theta} & 0 \\ 0 & e^{\pm i\theta} \end{pmatrix},$$

hence $\rho_{\pm\theta}$ is its own Jacobian. Since $\rho_{\pm\theta}$ is a multiple of the identity, it commutes with all other 2×2 matrices. So

$$\mathrm{Jac}_{\mathbb{C}} g(0) = \mathrm{id}.$$

Now the preceding proposition applies and we may conclude that

$$g(z) \equiv z.$$

This means that

$$f \circ \rho_\theta = \rho_\theta \circ f. \tag{$*$}$$

We now write f in a convergent power series in a neighborhood of 0:

$$f(z) = \sum a_{j,k} z_1^j z_2^k.$$

Then $(*)$ says that

$$\sum a_{j,k} e^{i(j+k)\theta} z_1^j z_2^k = \sum a_{j,k} e^{i\theta} z_1^j z_2^k$$

for all θ (where this is an equation of ordered pairs). Equating like monomials implies that

$$a_{j,k} = 0$$

unless $j + k = 1$. It follows that f is linear near 0. By the uniqueness of analytic continuation (Lemma 2), f is linear. ∎

Proposition 7. *Let* $\phi \in \text{Aut}(D^2(0, 1))$. *Then there exist* a_1, $a_2 \in D(0, 1)$, θ_1, $\theta_2 \in [0, 2\pi)$, *and a permutation* σ *of the set* $\{1, 2\}$ *such that*

$$\phi(z) = \left(e^{i\theta_1} \cdot \frac{z_{\sigma(1)} - a_1}{1 - \bar{a}_1 z_{\sigma(1)}}, \; e^{i\theta_2} \cdot \frac{z_{\sigma(2)} - a_2}{1 - \bar{a}_2 z_{\sigma(2)}} \right).$$

Proof. Let $\phi(0) = \alpha = (\alpha_1, \alpha_2)$. Define

$$\psi(z) = \left(\frac{z_1 - \alpha_1}{1 - \bar{\alpha}_1 z_1}, \; \frac{z_2 - \alpha_2}{1 - \bar{\alpha}_2 z_2} \right).$$

Then $g \equiv \psi \circ \phi \in \text{Aut}(D^2(0, 1))$ and $g(0) = 0$. It suffices to show that

$$g(z_1, z_2) = \left(e^{i\theta_1} z_{\sigma(1)}, \; e^{i\theta_2} z_{\sigma(2)} \right).$$

We know from Proposition 6 that g is linear. So

$$g(z) = \begin{pmatrix} g_{11} & g_{12} \\ g_{21} & g_{22} \end{pmatrix} \begin{pmatrix} z_1 \\ z_2 \end{pmatrix}$$

for some complex constants g_{ij} of modulus not exceeding one. Define, for $k = 2, 3, \ldots$, points $z^{1,k}$, $z^{2,k} \in D^2(0, 1)$ by

$$z^{1,k} = \left((1 - 1/k)\overline{\text{sgn}(g_{11})}, \ (1 - 1/k)\overline{\text{sgn}(g_{12})}\right),$$

$$z^{2,k} = \left((1 - 1/k)\overline{\text{sgn}(g_{21})}, \ (1 - 1/k)\overline{\text{sgn}(g_{22})}\right).$$

Here, for $w \in \mathbb{C}$, we are letting

$$\text{sgn } w \equiv \begin{cases} w/|w|, & w \neq 0, \\ 0, & w = 0. \end{cases}$$

Now

$$g(z^{1,k}) \text{ has first component} (1 - 1/k)(|g_{11}| + |g_{12}|)$$

and

$$g(z^{2,k}) \text{ has second component} (1 - 1/k)(|g_{21}| + |g_{22}|).$$

Letting $k \to +\infty$ gives

$$|g_{11}| + |g_{12}| \leq 1$$
$$|g_{21}| + |g_{22}| \leq 1. \tag{$*$}$$

On the other hand, apply g to $\alpha^k = (1 - 1/k, 0)$ and $\beta^k = (0, 1 - 1/k)$ and let $k \to +\infty$. We conclude that

$$((1 - 1/k)g_{11}, \ (1 - 1/k)g_{21}) \longrightarrow \partial D^2(0, 1)$$

and

$$((1 - 1/k)g_{12}, \ (1 - 1/k)g_{22}) \longrightarrow \partial D^2(0, 1).$$

Therefore

$$\max\{|g_{11}|, \ |g_{21}|\} = 1$$
$$\max\{|g_{12}|, \ |g_{22}|\} = 1. \tag{$**$}$$

The only way that $(*)$ and $(**)$ can both be true is if each column of the matrix (g_{ij}) has one entry of modulus one and one entry 0. Say that $|g_{\eta(k),k}| = 1$ for $k = 1, 2$, where η is a permutation of $\{1, 2\}$. Letting $\sigma = \eta^{-1}$ gives that

$$|g_{j,\sigma(j)}| = 1, \qquad j = 1, 2$$

and

$$g_{j,k} = 0 \qquad \text{if} \quad k \neq \sigma(j).$$

Let $g_{j,\sigma(j)} = e^{i\theta_j}$, $j = 1, 2$. Then

$$g(z_1, z_2) = (e^{i\theta_1} \cdot z_{\sigma(1)}, \ e^{i\theta_2} \cdot z_{\sigma(2)}),$$

as desired. ∎

An elegant and complete treatment of the biholomorphic self-maps of the unit ball may be found in [RU2]. We give a more elementary treatment here.

Proposition 8. *If $a \in \mathbb{C}$, $|a| < 1$, then the map*

$$\phi_a(z_1, z_2) = \left(\frac{z_1 - a}{1 - \bar{a}z_1}, \ \frac{(1 - |a|^2)^{1/2}z_2}{1 - \bar{a}z_1} \right)$$

is a biholomorphic self-map of $\mathcal{B}(0, 1)$ to $\mathcal{B}(0, 1)$.

Remark. The maps ϕ_a are natural generalizations of the Möbius transformations which we used extensively in the first part of the book. ∎

Proof of the Proposition. Now

$$|\phi_a(z)| < 1$$

if and only if

$$\left| \frac{z_1 - a}{1 - \bar{a}z_1} \right|^2 + \left| \frac{(1 - |a|^2)^{1/2}z_2}{1 - \bar{a}z_1} \right|^2 < 1$$

if and only if

$$|z_1 - a|^2 + (1 - |a|^2)|z_2|^2 < |1 - \bar{a}z_1|^2$$

if and only if

$$|z_1|^2 + |a|^2 + (1 - |a|^2)|z_2|^2 < 1 + |a|^2|z_1|^2$$

if and only if

$$(1 - |a|^2)|z_1|^2 + (1 - |a|^2)|z_2|^2 < 1 - |a|^2$$

if and only if

$$z \in \mathcal{B}(0, 1).$$ ■

Exercise. The inverse of the mapping ϕ_a is ϕ_{-a}.   ■

Recall that a 2×2 matrix

$$M : \mathbb{C}^2 \longrightarrow \mathbb{C}^2$$

is *unitary* if its inverse equals its conjugate transpose. Geometrically, M is unitary if and only if it takes any orthonormal basis of $\mathbb{C}^2$ (over the base field $\mathbb{C}$) to another.

Proposition 9. *If $g : \mathcal{B}(0, 1) \to \mathcal{B}(0, 1)$ is biholomorphic and $g(0) = 0$, then g is a unitary rotation (i.e., a mapping induced by a unitary matrix).*

Proof. Since $\mathcal{B}(0, 1)$ is complete circular, the map g must be linear (by Proposition 6). Now g must preserve the boundary of $\mathcal{B}(0, 1)$. [Why?— Think about a linear map acting on a closed, convex set; it will preserve the extreme points.] Therefore the function g takes Euclidean unit vectors to Euclidean unit vectors. But this says that g is unitary.   ■

Now we can characterize the automorphisms of the unit ball.

Proposition 10. *If* $f : \mathcal{B}(0, 1) \to \mathcal{B}(0, 1)$ *is a biholomorphic mapping, then* f *is the composition of at most two unitary maps and a map of the form* ϕ_a.

Proof. Let $w = f(0)$. There is a unitary rotation α such that

$$\alpha(w) = (|w|, 0).$$

The map

$$g = \phi_{|w|} \circ \alpha \circ f$$

is therefore a biholomorphic map of $\mathcal{B}(0, 1)$ to $\mathcal{B}(0, 1)$ which takes 0 to 0. By Proposition 9, g is a unitary map. Call it β. Then

$$f = \alpha^{-1} \circ (\phi_{|w|})^{-1} \circ \beta.$$

The reader may verify that $(\phi_{|w|})^{-1} = \phi_{-|w|}$. The proof is therefore complete. $\blacksquare$

Exercise. The group $\mathrm{Aut}(D^2(0, 1))$ acts transitively on $D^2(0, 1)$. The group $\mathrm{Aut}(\mathcal{B}(0, 1))$ acts transitively on $\mathcal{B}(0, 1)$. $\blacksquare$

The ball and the bidisc are the two most fundamental domains in $\mathbb{C}^2$. Both are contractible, both have transitive automorphism groups, and both are domains of convergence for power series. It came therefore as a great surprise when Poincaré proved that there is no biholomorphic mapping

$$\phi : D^2(0, 1) \longrightarrow \mathcal{B}(0, 1).$$

He proved this result by first showing that if such a ϕ existed, then the induced map

$$\mathrm{Aut}(D^2(0, 1)) \ni h \longrightarrow \phi \circ h \circ \phi^{-1} \in \mathrm{Aut}(\mathcal{B}(0, 1))$$

would be a group isomorphism. He then analyzed $\mathrm{Aut}(D^2(0, 1))$ and $\mathrm{Aut}(\mathcal{B}(0, 1))$ *as groups* and showed that they could not be isomorphic.

Indeed, we may assume that ϕ maps 0 to 0. Then ϕ induces an isomorphism of the isotropy group of 0 (the collection of self-maps that fix 0) in the connected component of the identity in $\mathrm{Aut}(D^2(0,1))$ with the isotropy group of 0 in the connected component of the identity in $\mathrm{Aut}(\mathcal{B}(0,1))$. It is easy to see that the former is abelian while the latter, being the unitary group, is not. Thus these subgroups cannot be isomorphic and we have a contradiction. We have given a sufficiently detailed description of the automorphism groups that the interested reader may carry out Poincaré's program as an exercise.

We shall instead approach the biholomorphic equivalence problem using some of the geometric ideas developed in this monograph. The next section is devoted to that topic.

3. Invariant Metrics and the Inequivalence of the Ball and the Bidisc

We continue to use the symbol D to denote the unit disc in $\mathbb{C}$. Let $U \subseteq \mathbb{C}^2$ be a domain and $P \in U$. Define $(D, U)_P$ to be the holomorphic functions $f : U \to D$ such that $f(P) = 0$. Define $(U, D)^P$ to be the holomorphic functions $f : D \to U$ such that $f(0) = P$. We are now going to define the Carathéodory and Kobayashi metrics. Since we are working in a two-dimensional space, we can no longer specify a metric as a scalar-valued function on the domain. In fact a metric will measure the length of a vector at a point.

Definition 1. If $P \in U$ and $\xi \in \mathbb{C}^2$, then define the *Carathéodory length* of ξ at P to be

$$F_C^U(P, \xi) = \sup\{|\mathrm{Jac}_{\mathbb{C}} f(P) \cdot \xi| : f \in (D, U)_P\}.$$

Note here that $\mathrm{Jac}_{\mathbb{C}} f(P)$ is a 2-tuple of complex numbers (because it is the derivative of a mapping from two complex dimensions to one complex dimension) and so is ξ. Thus $\mathrm{Jac}_{\mathbb{C}} f(P) \cdot \xi$ is the dot product of two 2-tuples, and hence is a complex number.

Definition 2. If $P \in U$ and $\xi \in \mathbb{C}^2$, then define the *Kobayashi length* of ξ at P to be

$$F_K^U(P, \xi) = \inf\{|\xi|/|g'(0)| : g \in (U, D)^P,$$

$$g'(0) \text{ is a scalar multiple of } \xi\}.$$

Here and throughout, $|\ \ |$ denotes Euclidean length.

The way that we define metrics is still motivated by Riemann's philosophy. If $\gamma : [0, 1] \to U$ is a continuously differentiable curve, we define its Kobayashi length to be

$$\ell_K(\gamma) = \int_0^1 F_K^U(\gamma(t), \gamma'(t)) \, dt.$$

Notice that we are integrating the lengths, in the metric, of the tangent vectors to the curve. The Carathéodory length of a curve is defined similarly. We comment, just as in Section 3.1, that F_C and F_K are integrable because they are semicontinuous.

One of the basic properties which we shall prove is that holomorphic mappings decrease distance in the Carathéodory and Kobayashi metrics. In several variables we express this assertion as follows.

Proposition 3. *Let U_1 and U_2 be domains and*

$$f : U_1 \longrightarrow U_2$$

be a holomorphic mapping. If $P \in U_1$ and $\xi \in \mathbb{C}^2$, we define

$$f_*(P)\xi \equiv \text{Jac}_{\mathbb{C}} f(P) \cdot \xi.$$

Then

$$F_C^{U_1}(P, \xi) \geq F_C^{U_2}(f(P), f_*(P)\xi)$$

and

$$F_K^{U_1}(P, \xi) \geq F_K^{U_2}(f(P), f_*(P)\xi).$$

Proof. We give the proof for the Carathéodory metric. The proof for the Kobayashi metric is similar.

Choose $\phi \in (D, U_2)_{f(P)}$. Then $\phi \circ f \in (D, U_1)_P$. Hence

$$F_C^{U_1}(P, \xi) \geq |(\mathrm{Jac}_{\mathbb{C}}(\phi \circ f)(P))\xi|$$
$$= |\mathrm{Jac}_{\mathbb{C}}\phi(f(P)) \cdot (\mathrm{Jac}_{\mathbb{C}}f(P)) \cdot \xi|$$
$$= |\mathrm{Jac}_{\mathbb{C}}\phi(f(P)) \cdot (f_*(P))\xi|.$$

Taking the supremum over all ϕ gives

$$F_C^{U_1}(P, \xi) \geq F_C^{U_2}(f(P), f_*(P)\xi). \qquad \blacksquare$$

Corollary 3.1. *If f is a biholomorphic map, then f preserves both the Carathéodory and the Kobayashi metrics; that is, the inequalities in the Proposition become equalities.*

Proof. Obvious. Apply the proposition both to f and to f^{-1}. $\qquad \blacksquare$

Exercise. Check that the Proposition implies that f decreases the invariant lengths of curves. That is, if γ is a continuously differentiable curve in U_1 and $f_*\gamma \equiv f \circ \gamma$ is the corresponding curve in U_2, then show that $\ell_C(f_*\gamma) \leq \ell_C(\gamma)$ and $\ell_K(f_*\gamma) \leq \ell_K(\gamma)$. $\qquad \blacksquare$

We use Proposition 3 to define two interesting new invariants. These invariants were trivial in one complex variable, but now they provide crucial information.

Definition 4. Let $U \subseteq \mathbb{C}^2$ be a domain and $P \in U$. The *Carathéodory indicatrix* of U at P is

$$\mathbf{i}_P^C(U) = \{\xi \in \mathbb{C} : F_C^U(P, \xi) < 1\}.$$

The *Kobayashi indicatrix* of U at P is

$$[\mathbf{i}_P^K(U) = \{\xi \in \mathbb{C} : F_K^U(P, \xi) < 1\}.$$

In words, the indicatrix is the "unit ball," in the indicated metric, of vectors at P. Technically speaking, the indicatrix lives in the tangent space at P.

Proposition 5. *Let* $f : U_1 \rightarrow U_2$ *be a biholomorphic mapping of domains in* $\mathbb{C}^2$. *Say that* $f(P) = Q$. *Then*

$$\text{Jac}_{\mathbb{C}} f(P) : i_P^C(U_1) \longrightarrow i_Q^C(U_2)$$

and

$$\text{Jac}_{\mathbb{C}} f(P) : i_P^K(U_1) \longrightarrow i_Q^K(U_2)$$

are linear isomorphisms.

Proof. Since f is distance-decreasing in the Kobayashi metric, $\text{Jac}_{\mathbb{C}} f(P)$ maps $\mathbf{i}_P^K(U_1)$ into $\mathbf{i}_Q^K(U_2)$. But the same observation applies to

$$\text{Jac}_{\mathbb{C}}(f^{-1})(Q) = (\text{Jac}_{\mathbb{C}} f(P))^{-1};$$

it maps $\mathbf{i}_Q^K(U_2)$ into $\mathbf{i}_P^K(U_1)$. Thus $\text{Jac}_{\mathbb{C}} f(P)$ is a linear isomorphism of $\mathbf{i}_P^K$ to $\mathbf{i}_Q^K$ as claimed.

The proof for $\mathbf{i}_P^C$ is identical. ∎

Proposition 6. *Let* $B = \mathcal{B}(0, 1)$. *Then we have*

$$i_0^K(B) = B.$$

Proof. Let $\phi \in (B, D)^0$. If η is any Euclidean unit vector in $\mathbb{C}^2$ then consider the function

$$h(\zeta) \equiv \phi(\zeta) \cdot \eta.$$

Here "·" denotes the usual inner product of 2-vectors. We have that h maps the disc to the disc and $h(0) = 0$. By the Schwarz lemma of one variable,

$$|h'(0)| \leq 1.$$

Since this inequality holds for any choice of η, we conclude that

$$|\phi'(0)| \leq 1.$$

Now if ξ is any vector in $\mathbb{C}^2$ then it follows from the preceding calculation that

$$F_K^B(0, \xi) = \inf\{|\xi|/|\phi'(0)| : \phi \in (B, D)^0\}$$
$$\geq |\xi|.$$

On the other hand, the map

$$\phi_0(\zeta) \equiv \frac{\zeta}{|\xi|} \xi$$

satisfies $\phi_0 \in (B, D)^0$ and $\phi_0'(\zeta) = \xi/|\xi|$ for any ζ. Thus $\phi_0'(0) = (1/|\xi|) \cdot \xi$ is a positive multiple of ξ. Therefore

$$F_K^B(0, \xi) \leq |\xi|/|\phi_0'(0)| = |\xi|.$$

We conclude that

$$F_K^B(0, \xi) = |\xi|,$$

hence that

$$\mathbf{i}_0^K(B) = B. \qquad \blacksquare$$

Proposition 7. *Let $D^2 = D^2(0, 1)$. Then*

$$\mathbf{i}_0^K(D^2) = D^2.$$

Proof. Define the projections

$$\pi_1(z_1, z_2) = z_1 \qquad \text{and} \qquad \pi_2(z_1, z_2) = z_2.$$

Let $\eta = (\eta_1, \eta_2) \in \mathbb{C}^2$ be any vector. By Proposition 3, we have that

$$F_K^{D^2}(0, \eta) \geq F_K^{\pi_1(D^2)}((\pi_1(0), (\pi_1)_* \eta) = F_K^D(0, \eta_1).$$

But the Schwarz lemma of one variable tells us easily that the last quantity is just $|\eta_1|$. A similar argument shows that

$$F_K^{D^2}(0, \eta) \geq |\eta_2|.$$

We conclude from these two inequalities that

$$F_K^{D^2}(0, \eta) \geq \max\{|\eta_1|, |\eta_2|\}.$$

Therefore

$$\mathbf{i}_0^K(D^2) \subseteq D^2.$$

For the reverse inclusion, fix η as above and consider the function

$$\phi(\zeta) = \left(\frac{\zeta \eta_1}{\max\{|\eta_1|, |\eta_2|\}}, \frac{\zeta \eta_2}{\max\{|\eta_1|, |\eta_2|\}} \right), \quad \zeta \in D.$$

Then it is obvious that $\phi \in (D^2, D)^0$ and we see that

$$\phi'(\zeta) = \left(\frac{\eta_1}{\max\{|\eta_1|, |\eta_2|\}}, \frac{\eta_2}{\max\{|\eta_1|, |\eta_2|\}} \right), \quad \zeta \in D.$$

So $\phi'(0)$ is a positive multiple of η.
   Therefore

$$F_K^{D^2}(0, \eta) \leq \frac{|\eta|}{|\phi'(0)|} = \max\{|\eta_1|, |\eta_2|\}.$$

The opposite inclusion now follows.   ∎

Exercise. Verify that $\mathbf{i}_0^C(B) = B$ and $\mathbf{i}_0^C(D^2) = D^2$.   ∎

Theorem 8. (Poincaré). *There is no biholomorphic mapping of the bidisc D^2 to the ball B.*

Proof. Suppose that

$$\phi : D^2 \longrightarrow B$$

is biholomorphic. Let $\phi^{-1}(0) = \alpha \in D^2$. There is an element $\psi \in$ Aut$(D^2(0, 1))$ such that $\psi(0) = \alpha$. Consider $g \equiv \phi \circ \psi$. Then

$$g : D^2 \longrightarrow B$$

is biholomorphic and $g(0) = 0$. We will show that g cannot exist.

By Proposition 5, $\text{Jac}_{\mathbb{C}} g(0)$ is a linear isomorphism of $\mathbf{i}_0^K(D^2)$ to $\mathbf{i}_0^K(B)$. But Propositions 6 and 7 identify these as D^2 and B respectively. So we have that

$$\text{Jac}_{\mathbb{C}} g(0) : D^2 \longrightarrow B$$

is a linear isomorphism. However, this is impossible. For the segment $\mathbf{i} = \{(t + i0, 1) : 0 \leq t \leq 1\}$ lies in ∂D^2. The linear isomorphism would map $\mathbf{i}$ to a nontrivial segment in ∂B. But B is strictly convex (all boundary points are extreme) so its boundary contains no segments. This is the required contradiction.   ∎

Remark. It is important to appreciate the logic in this proof. The hypothesized map ϕ (and therefore g as well) is *not* assumed to extend in any way to ∂D^2. Indeed, given the very different natures of ∂D^2 and ∂B, we would expect ϕ to be highly pathological at the boundary. Our geometric machinery allows us to pass to the linear map $\text{Jac}_{\mathbb{C}} g(0)$, which is defined on all of space. Thus we are able to analyze the boundaries of the domains and arrive at a contradiction.   ∎

An important program begun by Poincaré in the early part of the twentieth century can be described as follows: If $\phi : \Omega_1 \to \Omega_2$ is a biholomorphic mapping of smoothly bounded domains in $\mathbb{C}^2$, then suppose that ϕ, ϕ^{-1} continue smoothly to the boundaries of Ω_1, Ω_2, respectively, in such a way that the extended functions are diffeomorphisms of the closures of the domains. Using just dimension theory (the number of degrees of freedom in ϕ versus the number of degrees of freedom in specifying the two real $(2n - 1)$-dimensional boundaries), it can be shown that infinitely many algebraic relations must be satisfied by the derivatives of functions parametrizing $\partial\Omega_1$ and $\partial\Omega_2$. These relations give rise to invariants which can play a vital role in the biholomorphic classification of domains.

It is only since 1974 that a substantial beginning has been made in fleshing out Poincaré's program—see the discussion at the end of Section 4.5. At about the same time, Chern and Moser [CHM] and Tanaka [TAN] created systems for calculating the boundary differential invariants. In his monumental work [FEF2], Fefferman actually gives an effective procedure for calculating the Chern/Moser/Tanaka invariants on a large and important class of domains.

More recently, S. R. Bell [BEL] and others have come up with simpler proofs of Fefferman's theorem for a much larger class of domains. While it is still believed that a biholomorphic mapping of any two smoothly bounded domains must extend smoothly to their respective boundaries, we are far from being able to prove this assertion.

Epilogue

In complex analysis, geometric methods provide both a natural language for analyzing and recasting classical problems and also a rubric for posing new problems. The interaction between the classical and the modern techniques is both rich and rewarding.

Many facets of this symbiosis have yet to be explored. In particular, very little is known about explicitly calculating and estimating the differential invariants described in the present monograph. It is hoped that this book will spark some new interest in these matters.

Appendix on the Structure Equations and Curvature

1. Introduction

Here we give a brief presentation of the connection between the calculus notion of curvature (see [THO]) and the more abstract notion of curvature which leads to the definition of κ in Chapter 2. It is a pleasure to acknowledge our debt to the clear and compelling exposition in [ONE].

First, a word about notation. We use the language of differential forms consistently in this appendix. On the one hand, classically trained analysts are often uncomfortable with this language. On the other hand, the best way to learn the language is to use it. And the context of curvature calculations on plane domains may in fact be the simplest non-trivial context in which differential forms can be profitably used. In any event, this appendix would be terribly clumsy if we did not use forms, so the decision is essentially automatic. All necessary background on differential forms may be found in [RU1] or in [ONE].

2. Expressing Curvature Intrinsically

First we recall the concept of curvature for a smooth, two-dimensional surface $M \subseteq \mathbb{R}^3$. All of our calculations are local, so it is convenient to think of M as parametrized by two coordinate functions over a connected open set $U \subseteq \mathbb{R}^2$:

$$U \ni (u, v) \xrightarrow{\ P\ } (x_1(u, v), x_2(u, v), x_3(u, v)) \in M.$$

We require that the matrix

$$\begin{pmatrix} \dfrac{\partial x_1}{\partial u} & \dfrac{\partial x_2}{\partial u} & \dfrac{\partial x_3}{\partial u} \\[3mm] \dfrac{\partial x_1}{\partial v} & \dfrac{\partial x_2}{\partial v} & \dfrac{\partial x_3}{\partial v} \end{pmatrix}$$

have rank 2 at each point of U. The vectors given by the rows of this matrix span the tangent plane to M at each point. Applying the Gram-Schmidt orthonormalization procedure to these row vectors, and shrinking U, M if necessary, we may create vector fields

$$E_1 : M \longrightarrow \mathbb{R}^3$$

$$E_2 : M \longrightarrow \mathbb{R}^3$$

such that $E_1(x_1, x_2, x_3)$ and $E_2(x_1, x_2, x_3)$ are orthonormal and tangent to M at each $P = (x_1, x_2, x_3) \in M$. Denote by $T_P(M)$ the collection of linear combinations

$$a E_1(x_1, x_2, x_3) + b E_2(x_1, x_2, x_3), \qquad a, b \in \mathbb{R}.$$

We call $T_P(M)$ the *tangent space* to M at P.

Let $E_3(P)$ be the unit normal to M at P given by the $E_1(P) \times E_2(P)$. The functions E_1, E_2, E_3 are smooth *vector fields* on M: they assign to each $P \in M$ a triple of orthonormal vectors.

Let $\delta_1, \delta_2, \delta_3$ denote the standard basis for vectors in $\mathbb{R}^3$:

$$\delta_1 = (1, 0, 0),$$

$$\delta_2 = (0, 1, 0),$$

$$\delta_3 = (0, 0, 1).$$

(Many calculus books call these vectors **i**, **j**, and **k**.) Then we may write

$$E_i = \sum_j a_{i,j}(x_1, x_2, x_3)\delta_j, \qquad i = 1, 2, 3.$$

The matrix

$$\mathcal{A} \equiv \left(a_{i,j}\right)_{i,j=1}^{3},$$

where the $a_{i,j}$ are *functions* of the space variables, is called the *attitude matrix* of the frame (or basis) E_1, E_2, E_3. Since $\mathcal{A}$ transforms one orthonormal frame to another, $\mathcal{A}$ is an orthogonal matrix. Hence

$$\mathcal{A}^{-1} = {}^{t}\mathcal{A}.$$

Definition 1. If $P \in M$, $v \in T_P(M)$, and f is a smooth function on M, then define

$$D_v f(P) = \frac{d}{dt} f \circ \phi(t) \bigg|_{t=0},$$

where ϕ is any smooth curve in M such that $\phi(0) = P$ and $\phi'(0) = v$. One checks that this definition is independent of the choice of ϕ.

Definition 2. If

$$\alpha : M \longrightarrow \mathbb{R}^3$$

is a vector field on M,

$$\alpha(P) = \alpha_1(P)\delta_1 + \alpha_2(P)\delta_2 + \alpha_3(P)\delta_3,$$

and $v \in T_P(M)$ then define

$$\nabla_v \alpha(P) = (D_v \alpha_1)(P)\delta_1 + (D_v \alpha_2)(P)\delta_2 + (D_v \alpha_3)(P)\delta_3.$$

The operation ∇_v is called *covariant differentiation* of the vector field α.

Definition 3. If $P \in M$, $v \in T_P(M)$, we define the *shape operator* (or *Weingarten map*) for M at P to be

$$S_P(v) = -\nabla_v E_3(P).$$

Lemma 4. *We have that $S_P(v) \in T_P(M)$.*

Proof. Now $E_3 \cdot E_3 \equiv 1$, hence

$$0 = D_v(E_3 \cdot E_3)\Big|_P$$

$$= (2 \nabla_v E_3) \cdot E_3\Big|_P$$

$$= -2 S_P(v) \cdot E_3(P).$$

Hence $S_P(v) \perp E_3(P)$ so $S_P(v) \in T_P(M)$.   ∎

Notice that the shape operator assigns to each $P \in M$ a linear operator S_P on the 2-dimensional tangent space $T_P(M)$.

We can express this linear operator as a matrix with respect to the basis $E_1(P), E_2(P)$. So S assigns to each $P \in M$ a 2×2 matrix $\mathcal{M}_P$.

The linear operator S_P measures the rate of change of E_3 in any tangent direction v. It can be shown that $\mathcal{M}_P$ is diagonalizable. The (real) eigenvectors of $\mathcal{M}_P$ correspond to the *principal curvatures* of M at P (these are the directions of greatest and least curvature) and the corresponding eigenvalues measure the amount of curvature in those directions.

The preceding observations about $\mathcal{M}_P$ motivate the following definition.

Definition 5. The *Gaussian curvature* $\kappa(P)$ of M at a point $P \in M$ is the determinant of $\mathcal{M}_P$—the product of the two eigenvalues.

Our aim is to express $\kappa(P)$ in terms of the intrinsic geometry of M, without reference to E_3—that is, without reference to the way that M is situated in space.

To this end, we define *covector fields* θ_i which are dual to the vector fields E_i:

$$\theta_i E_j(P) = \delta_{ij}.$$

Then the θ_i may be expressed as linear combinations of the standard basis covectors dx_1, dx_2, dx_3. Indeed, if $\mathcal{A} = (a_{i,j})$ is the attitude matrix then

$$\theta_i = \sum a_{i,j}\, dx_j$$

(remember that $\mathcal{A}^{-1} = {}^t\mathcal{A}$). Thus, for each i, θ_i is a differential form; and the standard calculus of differential forms—including exterior differentiation—applies.

From now on, we restrict attention to 1- and 2-forms acting on tangent vectors to M. Thus any 1-form α may be expressed as

$$\alpha = \alpha(E_1)\theta_1 + \alpha(E_2)\theta_2$$

and any 2-form β may be expressed as

$$\beta = \beta(E_1, E_2)\theta_1 \wedge \theta_2.$$

Now we define covector fields $\omega_{i,j}$, $i, j \in \{1, 2, 3\}$, by the formula

$$\omega_{i,j}(v) = (\nabla_v E_i) \cdot E_j(P).$$

Here "$\cdot$" is the Euclidean dot product. We think of $\omega_{i,j}$ as a differential 1-form. Notice that, since $E_i \cdot E_j \equiv \delta_{i,j}$, we have for $v \in T_P(M)$ that

$$\begin{aligned}
0 &= D_v(E_i \cdot E_j) \\
&= (\nabla_v E_i) \cdot E_j + E_i \cdot (\nabla_v E_j) \\
&= \omega_{i,j}(v) + \omega_{j,i}(v).
\end{aligned}$$

Thus

$$\omega_{i,j} = -\omega_{j,i}.$$

In particular,

$$\omega_{i,i} = 0.$$

If $v \in T_P(M)$ then it is easy to check, just using linear algebra, that

$$\nabla_v E_i = \sum_j \omega_{i,j}(v) E_j, \qquad 1 \le i \le 3.$$

We call the $\omega_{i,j}$ the *connection forms* for M. We can now express the shape operator in terms of these connection forms.

Proposition 6. *Let $P \in M$ and $v \in T_P(M)$. Then*

$$S_P(v) = \omega_{1,3}(v) E_1(P) + \omega_{2,3}(v) E_2(P).$$

Proof. We have that

$$S_P(v) = -\nabla_v E_3$$
$$= -\sum_{j=1}^{3} (\nabla_v E_3 \cdot E_j) E_j$$
$$= -\sum_{j=1}^{3} \omega_{3,j}(v) E_j$$
$$= \omega_{1,3}(v) E_1 + \omega_{2,3}(v) E_2,$$

since $\omega_{3,3} = 0$. ∎

Now we can express Gaussian curvature in terms of the $\omega_{i,j}$.

Proposition 7. *We have that*

$$\omega_{1,3} \wedge \omega_{2,3} = \kappa\, \theta_1 \wedge \theta_2.$$

Proof. We need to calculate $\mathcal{M}_P$ in terms of the $\omega_{i,j}$. Proposition 6 gives that

$$S_P(E_1) = \omega_{1,3}(E_1) E_1 + \omega_{2,3}(E_1) E_2$$

and

$$S_P(E_2) = \omega_{1,3}(E_2) E_1 + \omega_{2,3}(E_2) E_2$$

so the matrix of S_P, in terms of the basis E_1, E_2, is

$$\mathcal{M}_P = \begin{pmatrix} \omega_{1,3}(E_1) & \omega_{2,3}(E_1) \\ \omega_{1,3}(E_2) & \omega_{2,3}(E_2) \end{pmatrix}.$$

We know that $\omega_{1,3} \wedge \omega_{2,3}$, being a 2-form, can be written as $\lambda \cdot \theta_1 \wedge \theta_2$. On the other hand,

$$\begin{aligned} \kappa &= \det \mathcal{M}_P \\ &= \omega_{1,3}(E_1)\omega_{2,3}(E_2) - \omega_{1,3}(E_2)\omega_{2,3}(E_1) \\ &= (\omega_{1,3} \wedge \omega_{2,3})(E_1, E_2) \\ &= \lambda. \end{aligned}$$

Therefore

$$\omega_{1,3} \wedge \omega_{2,3} = \lambda\, \theta_1 \wedge \theta_2 = \kappa\, \theta_1 \wedge \theta_2. \qquad \blacksquare$$

Our goal now is to express κ using only those $\omega_{i,j}$ with $i \neq 3$, $j \neq 3$. For this we require a technical lemma about the attitude matrix.

Lemma 8. *We have that*

$$\omega_{i,j} = \sum_k a_{j,k}\, da_{i,k}, \qquad 1 \leq i,\ j \leq 3.$$

Proof. If $v \in T_P(M)$ then

$$\omega_{i,j}(v) = \nabla_v E_i \cdot E_j(P).$$

But

$$E_i = \sum_k a_{i,k} \delta_k.$$

Therefore

$$\nabla_v E_i = \sum_k (D_v a_{i,k}) \delta_k.$$

Then

$$\omega_{i,j}(v) \equiv \nabla_v E_i \cdot E_j$$

$$= \left(\sum_k (D_v a_{i,k}) \delta_k \right) \cdot \left(\sum_k a_{j,k} \delta_k \right)$$

$$= \sum_k (D_v a_{i,k}) a_{j,k}$$

$$= \sum_k da_{i,k}(v) a_{j,k}.$$

As a result,

$$\omega_{i,j} = \sum_k a_{j,k} \, da_{i,k}. \qquad \blacksquare$$

Now we have reached a milestone. We can derive the important *Cartan structural equations*, which are the key to our intrinsic formulas for curvature.

Theorem 9. *We have*

$$d\theta_i = \sum_j \omega_{i,j} \wedge \theta_j, \tag{1}$$

$$d\omega_{i,j} = \sum_k \omega_{i,k} \wedge \omega_{k,j}. \tag{2}$$

Proof. We have that

$$\theta_i = \sum_j a_{i,j} \, dx_j$$

hence

$$d\theta_i = \sum_j da_{i,j} \wedge dx_j.$$

Since the attitude matrix $\mathcal{A}$ is orthogonal, we may solve the equations in Lemma 8 for $da_{i,k}$ in terms of $\omega_{i,j}$. Thus

$$da_{i,j} = \sum_k \omega_{i,k} a_{k,j}.$$

Substituting this into our last formula gives

$$
\begin{aligned}
d\theta_i &= \sum_j \left[\left(\sum_k \omega_{i,k} a_{k,j} \right) \wedge dx_j \right] \\
&= \sum_k \left[\omega_{i,k} \wedge \sum_j a_{k,j}\, dx_j \right] \\
&= \sum_k \omega_{i,k} \wedge \theta_k.
\end{aligned}
$$

This is the first structural equation.

For the second equation, notice that the formula

$$\omega_{i,j} = \sum_k a_{j,k}\, da_{i,k}$$

implies

$$d\omega_{i,j} = -\sum_k da_{i,k} \wedge da_{j,k}.$$

On the other hand,

$$
\begin{aligned}
\sum_k \omega_{i,k} \wedge \omega_{k,j} &= \sum_k \left(\sum_\ell a_{k,\ell}\, da_{i,\ell} \right) \wedge \left(\sum_m a_{j,m}\, da_{k,m} \right) \\
&= \sum_k \left(\sum_\ell a_{k,\ell}\, da_{i,\ell} \right) \wedge \left(-\sum_m a_{k,m}\, da_{j,m} \right) \\
&= -\left(\sum_{\ell,m} \left\{ \sum_k a_{k,\ell} a_{k,m} \right\} \cdot da_{i,\ell} \wedge da_{j,m} \right)
\end{aligned}
$$

$$= -\sum_m da_{i,m} \wedge da_{j,m}$$

$$= d\omega_{i,j}.$$

In the antepenultimate line we have used the fact that $\mathcal{A}^{-1} = {}^t\mathcal{A}$. The result now follows.   ∎

The following corollary will prove critical.

Corollary. *We have*

$$d\omega_{1,2} = -\kappa\theta_1 \wedge \theta_2.$$

Proof. The second structural equation gives

$$d\omega_{1,2} = \sum_k \omega_{1,k} \wedge \omega_{k,2}$$

$$= \omega_{1,1} \wedge \omega_{1,2} + \omega_{1,2} \wedge \omega_{2,2} + \omega_{1,3} \wedge \omega_{3,2}.$$

Only the third summand doesn't vanish. But we calculated in Proposition 7 that it equals $-\kappa\theta_1 \wedge \theta_2$.   ∎

The Corollary has been our main goal in this subsection. It gives an intrinsic way to calculate Gaussian curvature in the classical setting, hence a way to *define* Gaussian curvature in more abstract settings. We now proceed to develop this more abstract point of view.

3. Curvature Calculations on Planar Domains

Let $\Omega \subseteq \mathbb{C}$ be a domain which is equipped with a metric ρ. Assume for simplicity that $\rho(z) > 0$ at all points of Ω. Define functions

$$E_1 \equiv \frac{(1,0)}{\rho} \qquad \text{and} \qquad E_2 \equiv \frac{(0,1)}{\rho}.$$

Then

$$\theta_1 = \rho\, dx \qquad \text{and} \qquad \theta_2 = \rho\, dy$$

are the dual covector fields. We define $\omega_{i,j}$ according to the first structure equation:

$$d\theta_1 = \omega_{1,2} \wedge \theta_2,$$
$$d\theta_2 = \omega_{2,1} \wedge \theta_1.$$

We define Gaussian curvature according to the Corollary to Theorem 9 in the previous section:

$$d\omega_{1,2} = -\kappa\theta_1 \wedge \theta_2.$$

One can check that these definitions are independent of the choice of frame (or basis) E_1, E_2, but this is irrelevant for our purposes.

We conclude this Appendix by proving that the definition of curvature which we just elicited from the structural equations coincides with the one given in Section 2.1. First, by the way that we've defined θ_1 and θ_2, we have

$$\begin{aligned}
d\theta_1 &= d\rho \wedge dx \\
&= (\rho_x\, dx + \rho_y\, dy) \wedge dx \\
&= \rho_y\, dy \wedge dx \\
&= -\frac{\rho_y}{\rho}\, dx \wedge \rho\, dy \\
&= -\frac{\rho_y}{\rho}\, dx \wedge \theta_2.
\end{aligned}$$

Similarly,

$$\begin{aligned}
d\theta_2 &= d\rho \wedge dy \\
&= (\rho_x\, dx + \rho_y\, dy) \wedge dy \\
&= \rho_x\, dx \wedge dy
\end{aligned}$$

$$= -\frac{\rho_x}{\rho} \, dy \wedge \rho \, dx$$

$$= -\frac{\rho_x}{\rho} \, dy \wedge \theta_1.$$

Comparison with the first structural equation gives

$$\omega_{1,2} = -\frac{\rho_y}{\rho} \, dx + \tau \, dy$$

and

$$\omega_{1,2} = -\omega_{2,1} = -\left(-\frac{\rho_x}{\rho} \, dy \right) + \sigma \, dx,$$

for some unknown functions τ and σ.

The only way these equations can be consistent is if

$$\omega_{1,2} = -\frac{\rho_y}{\rho} \, dx + \frac{\rho_x}{\rho} \, dy.$$

Thus

$$d\omega_{1,2} = -\frac{\partial}{\partial y}\left(\frac{\rho_y}{\rho}\right) dy \wedge dx + \frac{\partial}{\partial x}\left(\frac{\rho_x}{\rho}\right) dx \wedge dy$$

$$= \left(-\frac{\rho_{yy}}{\rho} + \frac{\rho_y \rho_y}{\rho^2} \right) dy \wedge dx + \left(\frac{\rho_{xx}}{\rho} - \frac{\rho_x \rho_x}{\rho^2} \right) dx \wedge dy$$

$$= \frac{1}{\rho^2}\left(\rho \Delta\rho - (\rho_y)^2 - (\rho_x)^2 \right) dx \wedge dy$$

$$= \frac{1}{\rho^4}\left(\rho \Delta\rho - (\rho_y)^2 - (\rho_x)^2 \right) \theta_1 \wedge \theta_2.$$

The second structural equation now implies that

$$\kappa = -\frac{1}{\rho^4}(\rho \Delta\rho - |\nabla\rho|^2).$$

On the other hand, in Section 2.1 we defined

$$\kappa = -\frac{\Delta \log \rho}{\rho^2}.$$

We have

$$\frac{\partial}{\partial \bar{z}} \log \rho = \frac{1}{\rho} \frac{\partial \rho}{\partial \bar{z}}$$

and

$$\Delta \log \rho = 4 \frac{\partial^2}{\partial z \partial \bar{z}} \log \rho$$

$$= 4 \left(-\frac{1}{\rho^2} \frac{\partial \rho}{\partial z} \frac{\partial \rho}{\partial \bar{z}} + \frac{1}{\rho} \frac{\partial^2 \rho}{\partial z \partial \bar{z}} \right)$$

$$= -\frac{1}{\rho^2} \cdot |\nabla \rho|^2 + \frac{1}{\rho} \cdot \Delta \rho.$$

It follows that, according to the definition in Section 2.1,

$$\kappa = -\frac{1}{\rho^4} (\rho \Delta \rho - |\nabla \rho|^2).$$

Thus, as claimed, the definition of curvature from Chapter 2 is equal to that which arises from the structural equations.

Table of Symbols

Symbol	Page Number	Meaning
$\mathcal{A}$	193	the attitude matrix
$A^2(\Omega)$	138	the Bergman space
$A_{r,R}$	122	an annulus
$\text{Aut}(U)$	121, 179	the automorphism group of U
$[az+b]/[cz+d]$	14	linear fractional transformation
$\mathbf{B}(0,r)$	48	Poincaré metric ball
$B(P,R)$	128	metric ball
$\mathcal{B}(P,r)$	165	a ball in $\mathbb{C}^2$
$\mathbb{C}$	1	complex numbers
$\widehat{\mathbb{C}}$	82	the Riemann sphere
$\mathbf{C}$	62	arc of a circle
$\mathcal{C}$	168	domain of convergence
$\mathbb{C}_{0,1}$	76	the domain $\mathbb{C} \setminus \{0,1\}$
C^1	3	continuously differentiable
C^2	105	twice continuously differentiable
C^k	105	k times continuously differentiable
C^∞	106	infinitely differentiable
$\mathcal{C}_\Omega(P,Q)$	37	all piecewise continuously differentiable curves connecting P to Q
$c(P)$	111	center of curvature
C_R	53	points in D having Poincaré distance R from 0
$C(z,\zeta)$	140	the Cauchy kernel
D	2	unit disc
$\mathbf{d}$	111	internally tangent disc
$\triangle$	40	the Laplacian
d_ρ	47	the Poincaré distance
D_v	193	directional derivative
$\delta_1, \delta_2, \delta_3$	192	standard basis vectors
dA	138	area measure
$\det \text{Jac}\,\phi$	141	the Jacobian determinant
$D(P,r)$	2	open disc
$\overline{D}(P,r)$	2	closed disc

Symbol	Page Number	Meaning
$\partial D(P, r)$	2	boundary of disc
$D'(0, \epsilon)$	87	a punctured disc
$D(C(P), r_0)$	110	externally tangent disc
$D(C'(P), r_0)$	110	internally tangent disc
$D^2(P, r)$	165	a bidisc in $\mathbb{C}^2$
$\overline{D}^2(P, r)$	166	closure of a bidisc
$(D, U)_P$	90, 180	holomorphic functions f from U to D such that $f(P) = 0$
$\operatorname{dist}_C(z, \tau_p)$	120	Carathéodory distance of z to τ_p
$d_\rho(P, Q)$	37	ρ-metric distance of P to Q
$d_\rho(P, Q)$	47	Poincaré distance
$d_\sigma(z, w)$	83	spherical distance
$d\omega_{i,j} = \sum_k \omega_{i,k} \wedge \omega_{k,j}$ $d\theta_i = \sum_j \omega_{i,j} \wedge \theta_j$	198	Cartan structural equations
dV	139	volume measure
E_1, E_2	192	a frame (basis) of vector fields
E_3	192	the unit normal vector field
F	2	a holomorphic function
F'	2	complex derivative of F
$\mathcal{F}$	16	a normal family
$F_C^U(P)$	90, 180	Carathéodory metric on U at P
$F_K^U(P)$	94, 181	Kobayashi metric on U at P
$f_*\gamma$	43, 181	the push-forward of γ
f^*	42	the pullback mapping
$f^\#(z)$	85	the spherical derivative
f^n	58, 101	nth iterate of f
$\|f\|_{A^2(\Omega)}$	138	the Bergman norm
$\langle f, g\rangle_{A^2(\Omega)}$	139	the Bergman inner product
$\mathcal{G}$	81	a family of holomorphic functions
γ	3, 30, 31	a curve
$\dot{\gamma}$	30	derivative of γ
$\gamma_{P,Q}(t)$	50	curve of least Poincaré length connecting P to Q
$\Gamma_\alpha(P)$	118	Stolz region, non-tangential approach region

Symbol	Page Number	Meaning
$G(z, \zeta)$	148	the Green's function
id	122	the identity mapping
$\mathbf{i}(z)$	127	the identity map
$\mathbf{i}_P^C(U)$	182	Carathéodory indicatrix
$\mathbf{i}_P^K(U)$	182	Kobayashi indicatrix
$I(z)$	83	inversion mapping
$\oint_\gamma F(z)\,dz$	4	complex line integral
∞	86	point at infinity on the Riemann sphere
Jac ϕ	141, 172	the Jacobian matrix
K	17, 18	a compact set
$\kappa = \kappa(P)$	67	curvature of the metric ρ
$\kappa_{U,\rho}(z) = \kappa_\rho(z)$	67	curvature of the metric ρ
K_Ω	139, 140	the Bergman kernel for Ω
λ	30	a metric
ℓ	61, 62	a line in Euclidean geometry
$\ell(\gamma)$	30	length of γ
$\ell_\rho(\gamma)$	34	length of curve γ in metric ρ
$\ell_K(\gamma)$	181	Kobayashi length of the curve γ
$\mathcal{M}_\beta(P)$	119	metric approach region
$\mathcal{M}_P$	194	matrix of the shape operator
$\mu(z)$	76	metric of negative curvature on $\mathbb{C} \setminus \{0, 1\}$
$\nabla_v \alpha(P)$	193	covariant derivative
ν_P	108	unit outward normal at P
ν'_P	108	unit inward normal at P
$\omega_{i,j}$	195	connection forms
P	82	the north pole of the Riemann sphere
P_k	25	kth degree Taylor polynomial
$p(x, y)$	82	the stereographic projection map
$p(z)$	9	a polynomial
Ω	8	a domain
(Ω_1, ρ_1)	43	metric pair
$\partial/\partial z, \partial/\partial \bar{z}$	38	complex differential operators
$\phi_a(z_1, z_2)$	177	biholomorphic mapping of B
$\phi_a(\zeta)$	13	a Möbius transformation
$\{\phi_j\}$	145	an orthonormal system for $A^2(\Omega)$
$\psi(z, w)$	155	the pseudohyperbolic metric

Symbol	Page Number	Meaning
Ψ	129	a homotopy
$\mathbb{R}$	14	the real numbers
$r(P)$	111	radius of curvature
ρ	31	a metric (weight)
ρ	70	the Poincaré metric
$\rho(z)$	106	defining function
$\rho_\alpha^A(z)$	72	dilated, scaled Poincaré distance metric
ρ_C^Ω	103, 104	Carathéodory metric
ρ_E^Ω	103, 104	Euclidean metric
ρ_K^Ω	103, 104	Kobayashi metric
ρ_Ω	152	Bergman metric
$\rho_r(z)$	70	the dilated Poincaré metric
ρ_τ	14	a rotation
$\rho(z)|dz|$	34	a conformal metric
$S_P(v)$	193	shape operator or Weingarten map
σ	122	the reflection map
$\sigma(z)$	69	the spherical metric
σ_0	20	extremal function
$\sum_j a_j(z-P)^j$	2	power series expansion
$\sum a_{jk}(z_1 - P_1)^j(z_2 - P_2)^k$	166	power series expansion
T	109	tubular coordinate mapping
$T_P(M)$	192	the tangent space to M at P
τ	84	induced Euclidean metric on sphere
τ_p	120	inward normal segment at p
θ_i	194	covector fields
U	2, 8	a domain
$(U, D)^P$	93, 180	holomorphic functions f from D to U such that $f(0) = P$
U_0	81	the slit plane
U_w	164	slice of a domain in $\mathbb{C}^2$
U^z	165	slice of a domain in $\mathbb{C}^2$
W	108	tubular neighborhood
$\|\xi\|_{\rho,z}$	31	metric length of ξ
$|\xi|$	31	Euclidean length of ξ